Clustering for Data Mining

A Data Recovery Approach

Chapman & Hall/CRC

Computer Science and Data Analysis Series

The interface between the computer and statistical sciences is increasing, as each discipline seeks to harness the power and resources of the other. This series aims to foster the integration between the computer sciences and statistical, numerical, and probabilistic methods by publishing a broad range of reference works, textbooks, and handbooks.

Proposals for the series should be sent directly to one of the series editors above, or submitted to:

Chapman & Hall/CRC
23-25 Blades Court
London SW15 2NU
UK

Published Titles

Bayesian Artificial Intelligence
Kevin B. Korb and Ann E. Nicholson

Pattern Recognition Algorithms for Data Mining
Sankar K. Pal and Pabitra Mitra

Exploratory Data Analysis with MATLAB®
Wendy L. Martinez and Angel R. Martinez

Clustering for Data Mining: A Data Recovery Approach
Boris Mirkin

Correspondence Analysis and Data Coding with JAVA and R
Fionn Murtagh

R Graphics
Paul Murrell

Computer Science and Data Analysis Series

Clustering for Data Mining

A Data Recovery Approach

Boris Mirkin

Chapman & Hall/CRC
Taylor & Francis Group
Boca Raton London New York Singapore

Published in 2005 by
Chapman & Hall/CRC
Taylor & Francis Group
6000 Broken Sound Parkway NW, Suite 300
Boca Raton, FL 33487-2742

Printed in the United States of America on acid-free paper
10 9 8 7 6 5 4 3 2 1

International Standard Book Number-10: 1-58488-534-3 (Hardcover)
International Standard Book Number-13: 978-1-58488-534-4 (Hardcover)
Library of Congress Card Number 2005041421

Library of Congress Cataloging-in-Publication Data

Mirkin, B. G. (Boris Grigorévich)
Clustering for data mining : a data recovery approach / Boris Mirkin.
p. cm. -- (Computer science and data analysis series ; 3)
Includes bibliographical references and index.
ISBN 1-58488-534-3
1. Data mining. 2. Cluster analysis. I. Title. II. Series.

QA76.9.D343M57 2005
006.3'12--dc22 2005041421

T&F informa

Taylor & Francis Group
is the Academic Division of T&F Informa plc.

Visit the Taylor & Francis Web site at
http://www.taylorandfrancis.com

and the CRC Press Web site at
http://www.crcpress.com

Contents

Preface

Clustering is a discipline devoted to finding and describing cohesive or homogeneous chunks in data, the clusters.

Some exemplary clustering problems are:

- Finding common surf patterns in the set of web users;
- Automatically revealing meaningful parts in a digitalized image;
- Partition of a set of documents in groups by similarity of their contents;
- Visual display of the environmental similarity between regions on a country map;
- Monitoring socio-economic development of a system of settlements via a small number of representative settlements;
- Finding protein sequences in a database that are homologous to a query protein sequence;
- Finding anomalous patterns of gene expression data for diagnostic purposes;
- Producing a decision rule for separating potentially bad-debt credit applicants;
- Given a set of preferred vacation places, finding out what features of the places and vacationers attract each other;
- Classifying households according to their furniture purchasing patterns and finding groups' key characteristics to optimize furniture marketing and production.

Clustering is a key area in data mining and knowledge discovery, which are activities oriented towards finding non-trivial or hidden patterns in data collected in databases.

Earlier developments of clustering techniques have been associated, primarily, with three areas of research: factor analysis in psychology [55], numerical taxonomy in biology [122], and unsupervised learning in pattern recognition [21].

Technically speaking, the idea behind clustering is rather simple: introduce a measure of similarity between entities under consideration and combine similar entities into the same clusters while keeping dissimilar entities in different clusters. However, implementing this idea is less than straightforward.

First, too many similarity measures and clustering techniques have been invented with virtually no support to a non-specialist user in selecting among them. The trouble with this is that different similarity measures and/or clustering techniques may, and frequently do, lead to different results. Moreover, the same technique may also lead to different cluster solutions depending on the choice of parameters such as the initial setting or the number of clusters specified. On the other hand, some common data types, such as questionnaires with both quantitative and categorical features, have been left virtually without any substantiated similarity measure.

Second, use and interpretation of cluster structures may become an issue, especially when available data features are not straightforwardly related to the phenomenon under consideration. For instance, certain data on customers available at a bank, such as age and gender, typically are not very helpful in deciding whether to grant a customer a loan or not.

Specialists acknowledge peculiarities of the discipline of clustering. They understand that the clusters to be found in data may very well depend not on only the data but also on the user's goals and degree of granulation. They frequently consider clustering as art rather than science. Indeed, clustering has been dominated by learning from examples rather than theory based instructions. This is especially visible in texts written for inexperienced readers, such as [4], [28] and [115].

The general opinion among specialists is that clustering is a tool to be applied at the very beginning of investigation into the nature of a phenomenon under consideration, to view the data structure and then decide upon applying better suited methodologies. Another opinion of specialists is that methods for finding clusters as such should constitute the core of the discipline; related questions of data pre-processing, such as feature quantization and standardization, definition and computation of similarity, and post-processing, such as interpretation and association with other aspects of the phenomenon, should be left beyond the scope of the discipline because they are motivated by external considerations related to the substance of the phenomenon under investigation. I share the former opinion and argue the latter because it is at odds with the former: in the very first steps of knowledge discovery, substantive considerations are quite shaky, and it is unrealistic to expect that they alone could lead to properly solving the issues of pre- and post-processing.

Such a dissimilar opinion has led me to believe that the discovered clusters must be treated as an "ideal" representation of the data that could be used for recovering the original data back from the ideal format. This is the idea of the data recovery approach: not only use data for finding clusters but also use clusters for recovering the data. In a general situation, the data recovered from aggregate clusters cannot fit the original data exactly, which can be used for evaluation of the quality of clusters: the better the fit, the better the clusters. This perspective would also lead to the addressing of issues in pre- and post-processing, which now becomes possible because parts of the data that are explained by clusters can be separated from those that are not.

The data recovery approach is common in more traditional data mining and statistics areas such as regression, analysis of variance and factor analysis, where it works, to a great extent, due to the Pythagorean decomposition of the data scatter into "explained" and "unexplained" parts. Why not try the same approach in clustering?

In this book, two of the most popular clustering techniques, K-Means for partitioning and Ward's method for hierarchical clustering, are presented in the framework of the data recovery approach. The selection is by no means random: these two methods are well suited because they are based on statistical thinking related to and inspired by the data recovery approach, they minimize the overall within cluster variance of data. This seems to be the reason of the popularity of these methods. However, the traditional focus of research on computational and experimental aspects rather than theoretical ones has contributed to the lack of understanding of clustering methods in general and these two in particular. For instance, no firm relation between these two methods has been established so far, in spite of the fact that they share the same square error criterion.

I have found such a relation, in the format of a Pythagorean decomposition of the data scatter into parts explained and unexplained by the found cluster structure. It follows from the decomposition, quite unexpectedly, that it is the divisive clustering format, rather than the traditional agglomerative format, that better suits the Ward clustering criterion. The decomposition has led to a number of other observations that amount to a theoretical framework for the two methods. Moreover, the framework appears to be well suited for extensions of the methods to different data types such as mixed scale data including continuous, nominal and binary features. In addition, a bunch of both conventional and original interpretation aids have been derived for both partitioning and hierarchical clustering based on contributions of features and categories to clusters and splits. One more strain of clustering techniques, one-by-one clustering which is becoming increasingly popular, naturally emerges within the framework giving rise to intelligent versions of K-Means, mitigating the need for user-defined setting of the number of clusters and their hypothetical prototypes. Most importantly, the framework leads to a set of mathematically proven properties relating classical clustering with other clustering techniques such as conceptual clustering and graph theoretic clustering as well as with other data mining concepts such as decision trees and association in contingency data tables.

These are all presented in this book, which is oriented towards a reader interested in the technical aspects of data mining, be they a theoretician or a practitioner. The book is especially well suited for those who want to learn WHAT clustering is by learning not only HOW the techniques are applied but also WHY. In this way the reader receives knowledge which should allow him not only to apply the methods but also adapt, extend and modify them according to the reader's own ends.

This material is organized in five chapters presenting a unified theory along with computational, interpretational and practical issues of real-world data mining with clustering:
- What is clustering (Chapter 1);
- What is data (Chapter 2);
- What is K-Means (Chapter 3);
- What is Ward clustering (Chapter 4);
- What is the data recovery approach (Chapter 5).

But this is not the end of the story. Two more chapters follow. Chapter 6 presents some other clustering goals and methods such as SOM (self-organizing maps) and EM (expectation-maximization), as well as those for conceptual description of clusters. Chapter 7 takes on "big issues" of data mining: validity and reliability of clusters, missing data, options for data pre-processing and standardization, etc. When convenient, we indicate solutions to the issues following from the theory of the previous chapters. The Conclusion reviews the main points brought up by the data recovery approach to clustering and indicates potential for further developments.

This structure is intended, first, to introduce classical clustering methods and their extensions to modern tasks, according to the data recovery approach, without learning the theory (Chapters 1 through 4), then to describe the theory leading to these and related methods (Chapter 5) and, in addition, see a wider picture in which the theory is but a small part (Chapters 6 and 7).

In fact, my prime intention was to write a text on classical clustering, updated to issues of current interest in data mining such as processing mixed feature scales, incomplete clustering and conceptual interpretation. But then I realized that no such text can appear before the theory is described. When I started describing the theory, I found that there are holes in it, such as a lack of understanding of the relation between K-Means and the Ward method and in fact a lack of a theory for the Ward method at all, misconceptions in quantization of qualitative categories, and a lack of model based interpretation aids. This is how the current version has become a threefold creature oriented toward:

1. Giving an account of the data recovery approach to encompass partitioning, hierarchical and one-by-one clustering methods;

2. Presenting a coherent theory in clustering that addresses such issues as (a) relation between normalizing scales for categorical data and measuring association between categories and clustering, (b) contributions of various elements of cluster structures to data scatter and their use in interpreta-

tion, (c) relevant criteria and methods for clustering differently expressed data, etc.;

3. Providing a text in data mining for teaching and self-learning popular data mining techniques, especially K-Means partitioning and Ward agglomerative and divisive clustering, with emphases on mixed data pre-processing and interpretation aids in practical applications.

At present, there are two types of literature on clustering, one leaning towards providing general knowledge and the other giving more instruction. Books of the former type are Gordon [39] targeting readers with a degree of mathematical background and Everitt et al. [28] that does not require mathematical background. These include a great deal of methods and specific examples but leave rigorous data mining instruction beyond the prime contents. Publications of the latter type are Kaufman and Rousseeuw [62] and chapters in data mining books such as Dunham [23]. They contain selections of some techniques reported in an ad hoc manner, without any concern on relations between them, and provide detailed instruction on algorithms and their parameters.

This book combines features of both approaches. However, it does so in a rather distinct way. The book does contain a number of algorithms with detailed instructions and examples for their settings. But selection of methods is based on their fitting to the data recovery theory rather than just popularity. This leads to the covering of issues in pre- and post-processing matters that are usually left beyond instruction. The book does contain a general knowledge review, but it concerns more of issues rather than specific methods. In doing so, I had to clearly distinguish between four different perspectives: (a) statistics, (b) machine learning, (c) data mining, and (d) knowledge discovery, as those leading to different answers to the same questions. This text obviously pertains to the data mining and knowledge discovery perspectives, though the other two are also referred to, especially with regard to cluster validation.

The book assumes that the reader may have no mathematical background beyond high school: all necessary concepts are defined within the text. However, it does contain some technical stuff needed for shaping and explaining a technical theory. Thus it might be of help if the reader is acquainted with basic notions of calculus, statistics, matrix algebra, graph theory and logics.

To help the reader, the book conventionally includes a list of denotations, in the beginning, and a bibliography and index, in the end. Each individual chapter is preceded by a boxed set of goals and a dictionary of base words. Summarizing overviews are supplied to Chapters 3 through 7. Described methods are accompanied with numbered computational examples showing the working of the methods on relevant data sets from those presented in Chapter 1; there are 58 examples altogether. Computations have been carried out with self-made programs for MATLAB®, the technical computing tool developed by The MathWorks (see its Internet web site www.mathworks.com).

The material has been used in the teaching of data clustering and visualization to MSc CS students in several colleges across Europe. Based on these experiences, different teaching options can be suggested depending on the course objectives, time resources, and students' background.

If the main objective is teaching clustering methods and there are very few hours available, then it would be advisable to first pick up the material on generic K-Means in sections 3.1.1 and 3.1.2, and then review a couple of related methods such as PAM in section 6.1.2, iK-Means in 3.3.1, Ward agglomeration in 4.1 and division in 4.2.1, single linkage in 6.2.1 and SOM in 6.1.6. Given a little more time, a review of cluster validation techniques from 7.6 including examples in 3.3.2 should follow the methods. In a more relaxed regime, issues of interpretation should be brought forward as described in 3.4, 4.2.3, 6.3 and 7.2.

If the main objective is teaching data visualization, then the starting point should be the system of categories described in 1.1.5, followed by material related to these categories: bivariate analysis in section 2.2, regression in 5.1.2, principal component analysis (SVM decomposition) in 5.1.3, K-Means and iK-Means in Chapter 3, Self-organizing maps SOM in 6.1.6 and graph-theoretic structures in 6.2.

Acknowledgments

Too many people contributed to the approach and this book to list them all. However, I would like to mention those researchers whose support was important for channeling my research efforts: Dr. E. Braverman, Dr. V. Vapnik, Prof. Y. Gavrilets, and Prof. S. Aivazian, in Russia; Prof. F. Roberts, Prof. F. McMorris, Prof. P. Arabie, Prof. T. Krauze, and Prof. D. Fisher, in the USA; Prof. E. Diday, Prof. L. Lebart and Prof. B. Burtschy, in France; Prof. H.-H. Bock, Dr. M. Vingron, and Dr. S. Suhai, in Germany. The structure and contents of this book have been influenced by comments of Dr. I. Muchnik (Rutgers University, NJ, USA), Prof. M. Levin (Higher School of Economics, Moscow, Russia), Dr. S. Nascimento (University Nova, Lisbon, Portugal), and Prof. F. Murtagh (Royal Holloway, University of London, UK).

Author

Boris Mirkin is a Professor of Computer Science at the University of London UK. He develops methods for data mining in such areas as social surveys, bioinformatics and text analysis, and teaches computational intelligence and data visualization.

Dr. Mirkin first became known for his work on combinatorial models and methods for data analysis and their application in biological and social sciences. He has published monographs such as "Group Choice" (John Wiley & Sons, 1979) and "Graphs and Genes" (Springer-Verlag, 1984, with S. Rodin). Subsequently, Dr. Mirkin spent almost ten years doing research in scientific centers such as Ecole Nationale Suprieure des Tlcommunications (Paris, France), Deutsches Krebs Forschnung Zentrum (Heidelberg, Germany), and Center for Discrete Mathematics and Theoretical Computer Science DIMACS, Rutgers University (Piscataway, NJ, USA). Building on these experiences, he developed a unified framework for clustering as a data recovery discipline.

List of Denotations

I	Entity set
N	Number of entities
V	Feature set
V_l	Set of categories of a categorical feature l
M	Number of column features
$X = (x_{iv})$	Raw entity-to-feature data table
$Y = (y_{iv})$	Standardized entity-to-feature data table; $y_{iv} = (x_{iv} - a_v)/b_v$ where a_v and b_v denote the shift and scale coefficients, respectively
$y_i = (y_{iv})$	M-dimensional vector corresponding to entity $i \in I$ according to data table Y
$y_i = (y_{iv})$	M-dimensional vector corresponding to entity $i \in I$ according to data table Y
(x, y)	Inner product of two vector points $x = (x_j)$ and $y = (y_j)$, $(x, y) = \sum_j x_j y_j$
$d(x, y)$	Distance (Euclidean squared) between two vector points $x = (x_j)$ and $y = (y_j)$, $d(x, y) = \sum_j (x_j - y_j)^2$
$\{S_1, ..., S_K\}$	Partition of set I in K disjoint classes $S_k \subset I$, $k = 1, ..., K$
K	Number of classes/clusters in a partition $S = \{S_1, ..., S_K\}$ of set I
N_k	Number of entities in class S_k of partition S $(k = 1, ..., K)$
$c_k = (c_{kv})$	Centroid of cluster S_k, $c_{kv} = \sum_{i \in S_k} y_{iv}/N_k$, $v \in V$
S_w, S_{w1}, S_{w2}	Parent-children triple in a cluster hierarchy, $S_w = S_{w1} \cup S_{w2}$
$dw(S_{w1}, S_{w2})$	Ward distance between clusters S_{w1}, with centroid c_1, and S_{w2}, with centroid c_2, $dw(S_{w1}, S_{w2}) = \frac{N_{w1}N_{w2}}{N_{w1}+N_{w2}} d(c_{w1}, c_{w2})$
N_{kv}	Number of entities in class S_k of partition S $(k = 1, ..., K)$ that fall in category $v \in V$; an entry in the contingency table between partition S and categorical feature l with set of categories V_l

N_{k+}	Marginal distribution: Number of entities in class S_k of partition S $(k = 1, ..., K)$ as related to a contingency table between partition S and another categorical feature
N_{+v}	Marginal distribution: Number of entities falling in category $v \in V_l$ of categorical feature l as related to a contingency table between partition S and categorical feature l
p_{kv}	Frequency N_{kv}/N
p_{k+}	N_{k+}/N
p_{+v}	N_{+v}/N
q_{kv}	Relative Quetelet coefficient, $q_{kv} = \frac{p_{kv}}{p_{k+}p_{+v}} - 1$
$T(Y)$	Data scatter, $T(Y) = \sum_{i \in I} \sum_{v \in V} y_{iv}^2$
$W(S_k, c_k)$	Cluster's square error, $W(S_k, c_k) = \sum_{i \in S_k} d(y_i, c_k)$
$W(S, c)$	K-Means square error criterion equal to the sum of $W(S_k, c_k)$, $k = 1, ...K$
$\beta(i, S_k)$	Attraction of $i \in I$ to cluster S_k

Introduction: Historical Remarks

Clustering is a discipline aimed at revealing groups, or clusters, of similar entities in data. The existence of clustering activities can be traced a hundred years back, in different disciplines in different countries.

One of the first was the discipline of ecology. A question the scientists were trying to address was of the territorial structure of the settlement of bird species and its determinants. They did field sampling to count numbers of various species at observation spots; similarity measures between spots were defined, and a method of analysis of the structure of similarity dubbed Wrozlaw taxonomy was developed in Poland between WWI and WWII (see publication of a later time [32]). This method survives, in an altered form, in diverse computational schemes such as single-linkage clustering and minimum spanning tree (see section 6.2.1).

Simultaneously, phenomenal activities in differential psychology initiated in the United Kingdom by the thrust of F. Galton (1822-1911) and supported by the mathematical genius of K. Pearson (1855-1936) in trying to prove that human talent is not a random gift but inherited, led to developing a body of multivariate statistics including the discipline of factor analysis (primarily, for measuring talent) and, as its offshoot, cluster analysis. Take, for example, a list of high school students and their marks at various disciplines such as maths, English, history, etc. If one believes that the marks are exterior manifestations of an inner quality, or factor, of talent, then one can assign a student i with a hidden factor score of his talent, z_i. Then marks x_{il} of student i at different disciplines l can be modeled, up to an error, by the product $c_l z_i$ so that $x_{il} \approx c_l z_i$ where factor c_l reflects the impact of the discipline l over students. The problem is to find the unknown z_i and c_l, given a set of students' marks over a set of disciplines. This was the idea behind a method proposed by K. Pearson in 1901 [106] that became the ground for later developments in Principal Component Analysis (PCA), see further explanation in section 5.1.3. To do the job of measuring hidden factors, F. Galton hired C. Spearman who developed a rather distinct method for factor analysis based on the assumption that no unique talent can explain various human abilities, but there are different, and independent, dimensions of talent such as linguistic or spatial ones. Each of these hidden dimensions must be presented by a corresponding independent factor so that the mark can be thought of as the total of factor scores weighted by their loadings. This idea proved fruitful in developing various personality theories and related psychological tests. However, methods for factor analysis developed between WWI and WWII were computationally intensive since they used the operation of inversion of a matrix of discipline-to-discipline similarity coefficients (covariances, to be exact). The operation of matrix inversion still can be a challenging task when the matrix size grows into thousands, and it was a nightmare before the electronic computer era even with a matrix size of a dozen. It was noted then that variables (in this case, disciplines) related to the same factor are highly correlated among themselves, which led to the idea of catching "clusters" of highly correlated variables as proxies for factors, without computing the inverse matrix, an activity which was referred to once as "factor analysis for the poor." The very first book on cluster analysis, within this framework, was published in 1939 [131], see also [55].

In the 50s and 60s of the 20th century, with computer powers made available at universities, cluster analysis research grew fast in many disciplines simultaneously. Three of these seem especially important for the development of cluster analysis as a scientific discipline.

First, machine learning of groups of entities (pattern recognition) sprang up to involve both supervised and unsupervised learning, the latter being synonymous to cluster analysis [21].

Second, the discipline of numerical taxonomy emerged in biology claiming that a biological taxon, as a rule, could not be defined in the Aristotelian way, with a conjunction of features: a taxon thus was supposed to be such a set of organisms in which a majority shared a majority of attributes with each other [122]. Hierarchical agglomerative and divisive clustering algorithms were supposed to formalize this. They were being "polythetic" by the very mechanism of their action in contrast to classical "monothetic" approaches in which every divergence of taxa was to be explained by a single character. (It should be noted that the appeal of numerical taxonomists left some biologists unimpressed; there even exists the so-called "cladistics" discipline that claims that a single feature ought always to be responsible for any evolutionary divergence.)

Third, in the social sciences, an opposite stance of building a divisive decision tree at which every split is made over a single feature emerged in the work of Sonquist and Morgan (see a later reference [124]). This work led to the development of decision tree techniques that became a highly popular part of machine learning and data mining. Decision trees actually cover three methods, conceptual clustering, classification trees and regression trees, that are usually considered different because they employ different criteria of homogeneity [58]. In a conceptual clustering tree, split parts must be as homogeneous as possible with regard to all participating features. In contrast, a classification tree or regression tree achieves homogeneity with regard to only one, so-called target, feature. Still, we consider that all these techniques belong in cluster analysis because they all produce split parts consisting of similar entities; however, this does not prevent them also being part of other disciplines such as machine learning or pattern recognition.

A number of books reflecting these developments were published in the 70s describing the great opportunities opened in many areas of human activity by algorithms for finding "coherent" clusters in a data "cloud" placed in geometrical space (see, for example, Benzécri 1973, Bock 1974, Clifford and Stephenson 1975, Duda and Hart 1973, Duran and Odell 1974, Everitt 1974, Hartigan 1975, Sneath and Sokal 1973, Sonquist, Baker, and Morgan 1973, Van Ryzin 1977, Zagoruyko 1972). In the next decade, some of these developments have been further advanced and presented in such books as Breiman et al. [11], Jain and Dubes [58] and McLachlan and Basford [82]. Still the common view is that clustering is an art rather than a science because determining clusters may depend more on the user's goals than on a theory. Accordingly, clustering is viewed as a set of diverse and ad hoc procedures rather than a consistent theory.

The last decade saw the emergence of data mining, the discipline combining issues of handling and maintaining data with approaches from statistics and machine learning for discovering patterns in data. In contrast to the statistical approach, which tries to find and fit objective regularities in data, data mining is oriented towards the end user. That means that data mining considers the problem of useful knowledge discovery in its entire range, starting from database acquisition to data preprocessing to finding patterns to drawing conclusions. In particular, the concept of an interesting pattern as something which is unusual or far from normal or anomalous has been introduced into data mining [29]. Obviously, an anomalous cluster is one that is further away from the grand mean or any other point of reference – an approach which is adapted in this text.

A number of computer programs for carrying out data mining tasks, clustering included, have been successfully exploited, both in science and industry; a review of them can be found in [23]. There are a number of general purpose statistical packages which have made it through from earlier times: those with some cluster analysis applications such as SAS [119] and SPSS[42] or those entirely devoted to clustering such as CLUSTAN [140]. There are data mining tools which include clustering, such as Clementine [14]. Still, these programs are far from sufficient in advising a user on what method to select, how to pre-process data and, especially, what sense to make of the clusters.

Another feature of this more recent period is that a number of application areas have emerged in which clustering is a key issue. In many application areas that began much earlier – such as image analysis, machine vision or robot planning – clustering is a rather small part of a very complex task such that the quality of clustering does not much matter to the overall performance; as any reasonable heuristic would do, these areas do not require the discipline of clustering to theoretically develop and mature.

This is not so in Bio-informatics, the discipline which tries to make sense of interrelation between structure, function and evolution of biomolecular objects. Its primary entities, DNA and protein sequences, are complex enough to have their similarity modeled as homology, that is, inheritance from a common ancestor. More advanced structural data such as protein folds and their contact maps are being constantly added to existing depositories. Gene expression technologies add to this an invaluable next step - a wealth of data on biomolecular function. Clustering is one of the major tools in the analysis of bioinformatics data. The very nature of the problem here makes researchers see clustering as a tool not only for finding cohesive groupings in data but also for relating the aspects of structure, function and evolution to each other. In this way, clustering is more and more becoming part of an emerging area of computer classification. It models the major functions of classification in the sciences: the structuring of a phenomenon and associating its different aspects. (Though, in data mining, the term 'classification' is almost exclusively used in its partial meaning as merely a diagnostic tool.) Theoretical and practical research in clustering is thriving in this area.

Another area of booming clustering research is information retrieval and text document mining. With the growth of the Internet and the World Wide Web, text has become one of the most important mediums of mass communication. The terabytes of text that exist must be summarized effectively, which involves a great deal of clustering in such key stages as natural language processing, feature extraction, categorization, annotation and summarization. In author's view, clustering will become even more important as the systems for acquiring and understanding knowledge from texts evolve, which is likely to occur soon. There are already web sites providing web search results with clustering them according to automatically found key phrases (see, for instance, [134]).

This book is mostly devoted to explaining and extending two clustering techniques, K-Means for partitioning and Ward for hierarchical clustering. The choice is far from random. First, they present the most popular clustering formats, hierarchies and partitions, and can be extended to other interesting formats such as single clusters. Second, many other clustering and statistical techniques, such as conceptual clustering, self-organizing maps (SOM), and contingency association measures, appear to be closely related to these. Third, both methods involve the same criterion, the minimum within cluster variance, which can be treated within the same theoretical framework. Fourth, many data mining issues of current interest, such as analysis of mixed data, incomplete clustering, and conceptual description of clusters, can be treated with extended versions of these methods. In fact, the book contents go far beyond these methods: the two last chapters, accounting for one third of the material, are devoted to the "big issues" in clustering and data mining that are not limited to specific methods.

The present account of the methods is based on a specific approach to cluster analysis, which can be referred to as the *data recovery clustering*. In this approach, clusters are not only found in data but they also feed back into the data: a cluster structure is used to generate data in the format of the data table which has been analyzed with clustering. The data generated by a cluster structure are, in a sense, "ideal" as they reproduce only the cluster structure lying behind their generation. The observed data can then be considered a noisy version of the ideal cluster-generated data; the extent of noise can be measured by the difference between the ideal and observed data. The smaller the difference the better the fit. This idea is not particularly new; it is, in fact, the backbone of many quantitative methods of multivariate statistics, such as regression and factor analysis. Moreover, it has been applied in clustering from the very beginning; in particular, Ward [135] developed his method of agglomerative clustering with implicitly this view of data analysis. Some methods were consciously constructed along the data recovery approach: see, for instance, work of Hartigan [46] at which the single linkage method was developed to approximate the data with an ultrametric matrix, an ideal data type corresponding to a cluster hierarchy. Even more appealing in this capacity is a later work by Hartigan [47].

However, this approach has never been applied in full. The sheer idea, following from models presented in this book, that classical clustering is but a constrained analogue to the principal component model has not achieved any popularity so far, though it has been around for quite a while [89], [90]. The unifying capability of the data recovery clustering is grounded on convenient relations which exist between data approximation problems and geometrically explicit classical clustering. Firm mathematical relations found between different parts of cluster solutions and data lead not only to explanation of the classical algorithms but also to development of a number of other algorithms for both finding and describing clusters. Among the former, principal-component-like algorithms for finding anomalous clusters and divisive clustering should be pointed out. Among the latter, a set of simple but efficient interpretation tools, that are absent from the multiple programs implementing classical clustering methods, should be mentioned.

Chapter 1

What Is Clustering

After reading this chapter the reader will have a general understanding of:

1. What clustering is and its basic elements.
2. Clustering goals.
3. Quantitative and categorical features.
4. Main cluster structures: partition, hierarchy, and single cluster.
5. Different perspectives at clustering coming from statistics, machine learning, data mining, and knowledge discovery.

A set of small but real-world clustering problems will be presented.

Base words

Association Finding interrelations between different aspects of a phenomenon by matching cluster descriptions in the feature spaces corresponding to the aspects.

Classification An actual or ideal arrangement of entities under consideration in classes to shape and keep knowledge, capture the structure of phenomena, and relate different aspects of a phenomenon in question to each other. This term is also used in a narrow sense referring to any activities in assigning entities to prespecified classes.

Cluster A set of similar data entities found by a clustering algorithm.

Cluster representative An element of a cluster to represent its "typical" properties. This is used for cluster description in domains knowledge of which is poor.

Cluster structure A representation of an entity set I as a set of clusters that form either a partition of I or hierarchy on I or an incomplete clustering of I.

Cluster tendency A description of a cluster in terms of the average values of relevant features.

Clustering An activity of finding and/or describing cluster structures in a data set.

Clustering goal Types of problems of data analysis to which clustering can be applied: associating, structuring, describing, generalizing and visualizing.

Clustering criterion A formal definition or scoring function that can be used in computational algorithms for clustering.

Conceptual description A logical statement characterizing a cluster or cluster structure in terms of relevant features.

Data A set of entities characterized by values of quantitative or categorical features. Sometimes data may characterize relations between entities such as similarity coefficients or transaction flows.

Data mining perspective In data mining, clustering is a tool for finding patterns and regularities within the data.

Generalization Making general statements about data and, potentially, about the phenomenon the data relate to.

Knowledge discovery perspective In knowledge discovery, clustering is a tool for updating, correcting and extending the existing knowledge. In this regard, clustering is but empirical classification.

Machine learning perspective In machine learning, clustering is a tool for prediction.

Statistics perspective In statistics, clustering is a method to fit a prespecified probabilistic model of the data generating mechanism.

Structuring Representing data with a cluster structure.

Visualization Mapping data onto a known "ground" image such as the coordinate plane or a genealogy tree – in such a way that properties of the data are reflected in the structure of the ground image.

1.1 Exemplary problems

Clustering is a discipline devoted to revealing and describing homogeneous groups of entities, that is, clusters, in data sets. Why would one need this? Here is a list of potentially overlapping objectives for clustering.

1. **Structuring**, that is, representing data as a set of groups of similar objects.

2. **Description** of clusters in terms of features, not necessarily involved in finding the clusters.

3. **Association**, that is, finding interrelations between different aspects of a phenomenon by matching cluster descriptions in spaces corresponding to the aspects.

4. **Generalization**, that is, making general statements about data and, potentially, the phenomena the data relate to.

5. **Visualization**, that is, representing cluster structures as visual images.

These categories are not mutually exclusive, nor do they cover the entire range of clustering goals but rather reflect the author's opinion on the main applications of clustering. In the remainder of this section we provide real-world examples of data and the related clustering problems for each of these goals. For illustrative purposes, small data sets are used in order to provide the reader with the opportunity of directly observing further processing with the naked eye.

1.1.1 Structuring

Structuring is the main goal of many clustering applications, which is to find principal groups of entities in their specifics. The cluster structure of an entity set can be looked at through different glasses. One user may wish to aggregate the set in a system of nonoverlapping classes; another user may prefer to develop a taxonomy as a hierarchy of more and more abstract concepts; yet another user may wish to focus on a cluster of "core" entities considering the rest as merely a nuisance. These are conceptualized in different types of cluster structures, such as a partition, a hierarchy, or a single subset.

Market towns

Table 1.1 represents a small portion of a list of thirteen hundred English market towns characterized by the population and services provided in each listed in the following box.

Market town features:	
P	Population resident in 1991 Census
PS	Primary Schools
Do	Doctor Surgeries
Ho	Hospitals
Ba	Banks and Building Societies
SM	National Chain Supermarkets
Pe	Petrol Stations
DIY	Do-It-Yourself Shops
SP	Public Swimming Pools
PO	Post Offices
CA	Citizen's Advice Bureaux (cheap legal advice)
FM	Farmers' Markets

For the purposes of social monitoring, the set of all market towns should be partitioned into similarity clusters in such a way that a representative from each of the clusters may be utilized as a unit of observation. Those characteristics of the clusters that separate them from the others should be used to properly select representative towns.

As further computations will show, the numbers of services on average follow the town sizes, so that the found clusters can be described mainly in terms of the population size. This set, as well as the complete set of almost thirteen hundred English market towns, consists of seven clusters that can be described as belonging to four tiers of population: large towns of about 17-20,000 inhabitants, two clusters of medium sized towns (8-10,000 inhabitants), three clusters of small towns (about 5,000 inhabitants) and a cluster of very small settlements with about 2,500 inhabitants. The difference between clusters in the same population tier is caused by the presence or absence of some service features. For instance, each of the three small town clusters is characterized by the presence of a facility, which is absent in two others: a Farm market, a Hospital and a Swimming pool, respectively. The number of clusters is determined in the process of computations (see sections 3.3, 3.4.2).

This data set is analyzed on pp. 52, 56, 68, 92, 94, 97, 99, 100, 101, 108.

Primates and Human origin

In Table 1.2, the data on genetic distances between Human and three genera of great apes are presented; the Rhesus monkey is added as a distant relative to certify the starting divergence event. It is well established that humans diverged from a common ancestor with chimpanzees approximately 5 million years ago, after a divergence from other great apes. Let us see how compatible with this conclusion the results of cluster analysis are.

Table 1.1: **Market towns:** Market towns in the West Country, England.

Town	P	PS	Do	Ho	Ba	SM	Pe	DIY	SP	PO	CA	FM
Ashburton	3660	1	0	1	2	1	2	0	1	1	1	0
Bere Alston	2362	1	0	0	1	1	0	0	0	1	0	0
Bodmin	12553	5	2	1	6	3	5	1	1	2	1	0
Brixham	15865	7	3	1	5	5	3	0	2	5	1	0
Buckfastleigh	2786	2	1	0	1	2	2	0	1	1	1	1
Bugle/Stenalees	2695	2	0	0	0	0	1	0	0	2	0	0
Callington	3511	1	1	0	3	1	1	0	1	1	0	0
Dartmouth	5676	2	0	0	4	4	1	0	0	2	1	1
Falmouth	20297	6	4	1	11	3	2	0	1	9	1	0
Gunnislake	2236	2	1	0	1	0	1	0	0	3	0	0
Hayle	7034	4	0	1	2	2	2	0	0	2	1	0
Helston	8505	3	1	1	7	2	3	0	1	1	1	1
Horrabridge/Yel	3609	1	1	0	2	1	1	0	0	2	0	0
Ipplepen	2275	1	1	0	0	0	1	0	0	1	0	0
Ivybridge	9179	5	1	0	3	1	4	0	0	1	1	0
Kingsbridge	5258	2	1	1	7	1	2	0	0	1	1	1
Kingskerswell	3672	1	0	0	0	1	2	0	0	1	0	0
Launceston	6466	4	1	0	8	4	4	0	1	3	1	0
Liskeard	7044	2	2	2	6	2	3	0	1	2	2	0
Looe	5022	1	1	0	2	1	1	0	1	3	1	0
Lostwithiel	2452	2	1	0	2	0	1	0	0	1	0	1
Mevagissey	2272	1	1	0	1	0	0	0	0	1	0	0
Mullion	2040	1	0	0	2	0	1	0	0	1	0	0
Nanpean/Foxhole	2230	2	1	0	0	0	0	0	0	2	0	0
Newquay	17390	4	4	1	12	5	4	0	1	5	1	0
Newton Abbot	23801	13	4	1	13	4	7	1	1	7	2	0
Padstow	2460	1	0	0	3	0	0	0	0	1	1	0
Penryn	7027	3	1	0	2	4	1	0	0	3	1	0
Penzance	19709	10	4	1	12	7	5	1	1	7	2	0
Perranporth	2611	1	1	0	1	1	2	0	0	2	0	0
Porthleven	3123	1	0	0	1	1	0	0	0	1	0	0
Saltash	14139	4	2	1	4	2	3	1	1	3	1	0
South Brent	2087	1	1	0	1	1	0	0	0	1	0	0
St Agnes	2899	1	1	0	2	1	1	0	0	2	0	0
St Austell	21622	7	4	2	14	6	4	3	1	8	1	1
St Blazey/Par	8837	5	2	0	1	1	4	0	0	4	0	0
St Columb Major	2119	1	0	0	2	1	1	0	0	1	1	0
St Columb Road	2458	1	0	0	0	1	3	0	0	2	0	0
St Ives	10092	4	3	0	7	2	2	0	0	4	1	0
St Just	2092	1	0	0	2	1	1	0	0	1	0	0
Tavistock	10222	5	3	1	7	3	3	1	2	3	1	1
Torpoint	8238	2	3	0	3	2	1	0	0	2	1	0
Totnes	6929	2	1	1	7	2	1	0	1	4	0	1
Truro	18966	9	3	1	19	4	5	2	2	7	1	1
Wadebridge	5291	1	1	0	5	3	1	0	1	1	1	0

Table 1.2: **Primates:** Distances between four Primate species and Rhesus monkey.

Genus	Human	Chimpanzee	Gorilla	Orangutan
Chimpanzee	1.45			
Gorilla	1.51	1.57		
Orangutan	2.98	2.94	3.04	
Rhesus monkey	7.51	7.55	7.39	7.10

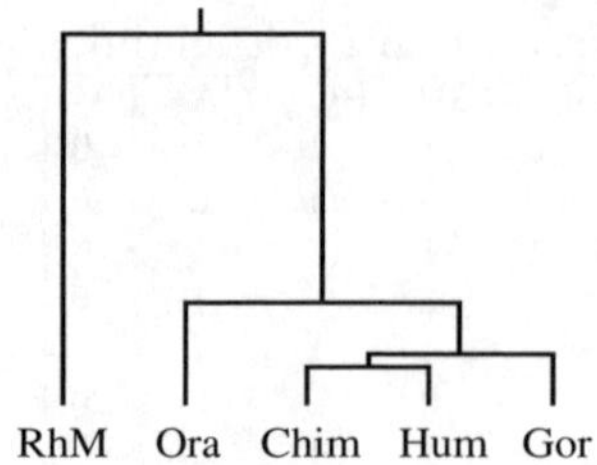

Figure 1.1: A tree representing pair-wise distances between the primate species from Table 1.2.

The data is a square matrix of the dissimilarity values between the species from Table 1.2 as cited in [90], p. 30. (Only sub-diagonal distances are shown since the table is symmetric.) An example of analysis of the structure of this matrix is given on p. 192.

The query: what species belongs to the same cluster as Humans? This obviously can be treated as a single cluster problem: one needs only one cluster to address the issue. The structure of the data is so simple that the cluster of chimpanzee, gorilla and human can be separated without any theory: distances within this subset are similar, all about the average 1.51, and by far less than other distances.

In biology, this problem is traditionally addressed through evolutionary trees, which are analogues to genealogy trees except that species play the role of relatives. An evolutionary tree built from the data in Table 1.2 is shown in Figure 1.1. The closest relationship between human and chimpanzee is obvious, with gorilla branching off next. The subject of human evolution is treated in depth with data mining methods in [13].

Gene presence-absence profiles

Evolutionary analysis is an important tool not only for understanding evolution but also for analysis of gene functions in humans and other organisms including medically and industrially important ones. The major assumption underlying the analysis is that all species are descendants of the same ancestor species, so that subsequent evolution can be depicted in terms of divergence only, as in the evolutionary tree in Figure 1.1.

The terminal nodes, so-called leaves, correspond to the species under consideration, and the root denotes the common ancestor. The other interior nodes represent other ancestral species, each being the last common ancestor to the set of organisms in the leaves of the sub-tree rooted in the given node. Recently, this line of research has been supplemented by data on the gene content of multiple species as exemplified in Table 1.3. Here, the columns correspond to 18 simple, unicellular organisms, bacteria and archaea (collectively called

Table 1.3: **Gene profiles**: Presence-absence profiles of 30 COGs in a set of 18 genomes.

No	COG	Species																	
		y	a	o	m	p	k	z	q	v	d	r	b	c	e	f	g	s	j
1	COG0090	1	1	1	1	1	1	1	1	1	1	1	1	1	1	1	1	1	1
2	COG0091	1	1	1	1	1	1	1	1	1	1	1	1	1	1	1	1	1	1
3	COG2511	0	1	1	1	1	1	1	0	0	0	0	0	0	0	0	0	0	0
4	COG0290	0	0	0	0	0	0	0	1	1	1	1	1	1	1	1	1	1	1
5	COG0215	1	1	1	0	1	1	1	1	1	1	1	1	1	1	1	1	1	1
6	COG2147	1	1	1	1	1	1	1	0	0	0	0	0	0	0	0	0	0	0
7	COG1746	0	1	1	1	1	1	1	0	0	0	0	0	0	0	0	0	0	0
8	COG1093	1	1	1	1	1	1	1	0	0	0	0	0	0	0	0	0	0	0
9	COG2263	0	1	1	1	1	1	1	0	0	0	0	0	0	0	0	0	0	0
10	COG0847	1	1	0	0	1	0	0	1	1	1	1	1	1	1	1	1	1	1
11	COG1599	1	1	1	1	1	1	1	0	0	0	0	0	0	0	0	0	0	0
12	COG3066	0	0	0	0	0	0	0	0	0	0	0	0	0	1	0	1	0	0
13	COG3293	0	0	0	0	0	0	0	0	0	1	1	0	1	0	0	0	0	1
14	COG3432	0	1	0	1	1	0	1	0	0	0	0	0	0	0	0	0	0	0
15	COG3620	0	1	0	0	1	1	0	0	0	0	0	0	0	0	0	0	0	0
16	COG1709	0	1	1	1	1	1	1	0	0	0	0	0	0	0	1	0	0	0
17	COG1405	1	1	1	1	1	1	1	0	0	0	0	0	0	0	0	0	0	0
18	COG3064	0	0	0	0	0	0	0	0	0	0	0	0	0	1	0	1	0	0
19	COG2853	0	0	0	0	0	0	0	0	0	0	0	0	0	1	1	1	1	1
20	COG2951	0	0	0	0	0	0	0	0	0	0	0	0	0	1	1	1	1	1
21	COG3114	0	0	0	0	0	0	0	0	0	0	0	0	0	1	1	1	0	0
22	COG3073	0	0	0	0	0	0	0	0	0	0	0	0	0	1	1	1	1	0
23	COG3026	0	0	0	0	0	0	0	0	0	0	0	0	0	1	1	1	0	0
24	COG3006	0	0	0	0	0	0	0	0	0	0	0	0	0	1	0	1	0	0
25	COG3115	0	0	0	0	0	0	0	0	0	0	0	0	0	1	1	1	1	0
26	COG2414	0	1	0	1	1	1	1	0	0	0	0	0	0	1	0	0	0	0
27	COG3029	0	0	0	0	0	0	0	0	0	0	1	0	0	1	0	1	0	0
28	COG3107	0	0	0	0	0	0	0	0	0	0	0	0	0	1	1	1	1	0
29	COG3429	0	0	0	0	0	0	0	0	0	1	1	0	1	0	0	0	0	0
30	COG1950	0	0	0	0	0	0	0	0	0	1	1	1	1	0	0	0	0	0

Table 1.4: **Species**: List of eighteen species (one eukaryota, then six archaea and then eleven bacteria) represented in Table 1.3.

Species	Code	Species	Code
Saccharomyces cerevisiae	y	*Deinococcus radiodurans*	d
Archaeoglobus fulgidus	a	*Mycobacterium tuberculosis*	r
Halobacterium sp.NRC-1	o	*Bacillus subtilis*	b
Methanococcus jannaschii	m	*Synechocystis*	c
Pyrococcus horikoshii	k	*Escherichia coli*	e
Thermoplasma acidophilum	p	*Pseudomonas aeruginosa*	f
Aeropyrum pernix	z	*Vibrio cholera*	g
Aquifex aeolicus	q	*Xylella fastidiosa*	s
Thermotoga maritima	v	*Caulobacter crescentus*	j

Table 1.5: COG names and functions.

Code	Name
COG0090	Ribosomal protein L2
COG0091	Ribosomal protein L22
COG2511	Archaeal Glu-tRNAGln
COG0290	Translation initiation factor IF3
COG0215	Cysteinyl-tRNA synthetase
COG2147	Ribosomal protein L19E
COG1746	tRNA nucleotidyltransferase (CCA-adding enzyme)
COG1093	Translation initiation factor eIF2alpha
COG2263	Predicted RNA methylase
COG0847	DNA polymerase III epsilon
COG1599	Replication factor A large subunit
COG3066	DNA mismatch repair protein
COG3293	Predicted transposase
COG3432	Predicted transcriptional regulator
COG3620	Predicted transcriptional regulator with C-terminal CBS domains
COG1709	Predicted transcriptional regulators
COG1405	Transcription initiation factor IIB
COG3064	Membrane protein involved
COG2853	Surface lipoprotein
COG2951	Membrane-bound lytic murein transglycosylase B
COG3114	Heme exporter protein D
COG3073	Negative regulator of sigma E
COG3026	Negative regulator of sigma E
COG3006	Uncharacterized protein involved in chromosome partitioning
COG3115	Cell division protein
COG2414	Aldehyde:ferredoxin oxidoreductase
COG3029	Fumarate reductase subunit C
COG3107	Putative lipoprotein
COG3429	Uncharacterized BCR, stimulates glucose-6-P dehydrogenase activity
COG1950	Predicted membrane protein

prokaryotes), and a simple eukaryote, yeast *Saccharomyces cerevisiae*. The list of species along with their one-letter codes is given in Table 1.4.

The rows in Table 1.3 correspond to individual genes represented by the so-called Clusters of Orthologous Groups (COGs) which are supposed to include genes originating from the same ancestral gene in the common ancestor of the respective species [68]. COG names which reflect the functions of the respective genes in the cell are given in Table 1.5. These tables present but a small part of the publicly available COG database currently including 66 species and 4857 COGs posted in the web site www.ncbi.nlm.nih.gov/COG.

The pattern of presence-absence of a COG in the analyzed species is shown in Table 1.3, with zeros and ones standing for absence and presence, respectively. This way, a COG can be considered a character (attribute) that is either present or absent in a species. Two of the COGs, in the top two rows, are present at each of the 18 genomes, whereas the others cover only some of the species.

An evolutionary tree must be consistent with the presence-absence patterns.

Specifically, if a COG is present in two species, then it should be present in their last common ancestor and, thus, in all other descendants of the last common ancestor. This would be in accord with the natural process of inheritance. However, in most cases, the presence-absence pattern of a COG in extant species is far from the "natural" one: many genes are dispersed over several subtrees. According to comparative genomics, this may happen because of multiple loss and horizontal transfer of genes [68]. The hierarchy should be constructed in such a way that the number of inconsistencies is minimized.

The so-called principle of Maximum Parsimony (MP) is a straightforward formalization of this idea. Unfortunately, MP does not always lead to appropriate solutions because of intrinsic and computational problems. A number of other approaches have been proposed including hierarchical cluster analysis (see [105]).

Especially appealing in this regard is divisive cluster analysis. It begins by splitting the entire data set into two parts, thus imitating the divergence of the last universal common ancestor (LUCA) into two descendants. The same process then applies to each of the split parts until a stop-criterion is reached to halt the division process. In contrast to other methods for building evolutionary trees, divisive clustering imitates the process of evolutionary divergence. Further approximation of the real evolutionary process can be achieved if the characters on which divergence is based are discarded immediately after the division of the respective cluster [96]. Gene profiles data are analyzed on p. 121 and p. 131.

After an evolutionary tree is built, it can be utilized for reconstructing gene histories by mapping events of emergence, inheritance, loss and horizontal transfer of individual COGs on the tree according to the principle of Maximum Parsimony (see p. 126). These histories of individual genes can be helpful in advancing our understanding of biological functions and drug design.

1.1.2 Description

The problem of description is that of automatically deriving a conceptual description of clusters found by a clustering algorithm or supplied from a different source. The problem of cluster description belongs in cluster analysis because this is part of the interpretation and understanding of clusters. A good conceptual description can be used for better understanding and/or better predicting. The latter because we can check whether an object in question satisfies the description or not: the more the object satisfies the description the better the chances that it belongs to the cluster described. This is why conceptual description tools, such as decision trees [11, 23], have been conveniently used and developed mostly for the purposes of prediction.

Describing Iris genera

Table 1.6 presents probably the most popular data set in the machine learning research community: 150 Iris specimens, each measured on four morphological variables: sepal length (w1), sepal width (w2), petal length (w3), and petal width (w4), as collected by botanist E. Anderson and published in a founding paper of celebrated British statistician R. Fisher in 1936 [7]. It is said that there are three species in the table, I *Iris setosa* (diploid), II *Iris versicolor* (tetraploid), and III *Iris virginica* (hexaploid), each represented by 50 consecutive entities in the corresponding column.

The classes are defined by the genome (genotype); the features are of the appearance (phenotype). Can the classes be described in terms of the features in Table 1.6? It is well known from previous studies that classes II and III are not well separated in the variable space (for example, specimens 28, 33 and 44 from class II are more similar to specimens 18, 26, and 33 from class III than to specimens of the same species, see Figure 1.10 on p. 25). This leads to the problem of deriving new features from those that have been measured on spot to provide for better descriptions of the classes. These new features could be then utilized for the clustering of additional specimens.

Some non-linear machine learning techniques such as Neural Nets [51] and Support Vector Machines [128] can tackle the problem and produce a decent decision rule involving non-linear transformation of the features. Unfortunately, rules that can be derived with currently available methods are not comprehensible to the human mind and, thus, cannot be used for interpretation and description. The human mind needs somewhat less artificial logics that can reproduce and extend such botanists' observations as that the petal area roughly expressed by the product of w3 and w4 provides for much better resolution than the original linear sizes. A method for building cluster descriptions of this type, referred to as APPCOD, will be described in section 7.2.

The Iris data set is analyzed on pp. 87, 211, 212, 213.

Body mass

Table 1.7 presents data on the height and weight of 22 males of which individuals p13-p22 are considered overweight and p1-p12 normal. As Figure 1.2 clearly shows, a line of best fit separating these two sets should run along the elongated cloud formed by entity points. The groups have been defined according to the so-called body mass index, bmi: those individuals whose bmi is 25 or over are considered overweight. The body mass index is defined as the ratio of the weight, in kilograms, to the squared height, in meters. The problem is to make a computer automatically transform the current height-weight feature space into such a format that would allow one to clearly distinguish between the overweight and normally-built individuals.

Table 1.6: **Iris:** Anderson-Fisher data on 150 Iris specimens.

Entity in a Class	Class I Iris setosa				Class II Iris versicolor				Class III Iris virginica			
	w1	w2	w3	w4	w1	w2	w3	w4	w1	w2	w3	w4
1	5.1	3.5	1.4	0.3	6.4	3.2	4.5	1.5	6.3	3.3	6.0	2.5
2	4.4	3.2	1.3	0.2	5.5	2.4	3.8	1.1	6.7	3.3	5.7	2.1
3	4.4	3.0	1.3	0.2	5.7	2.9	4.2	1.3	7.2	3.6	6.1	2.5
4	5.0	3.5	1.6	0.6	5.7	3.0	4.2	1.2	7.7	3.8	6.7	2.2
5	5.1	3.8	1.6	0.2	5.6	2.9	3.6	1.3	7.2	3.0	5.8	1.6
6	4.9	3.1	1.5	0.2	7.0	3.2	4.7	1.4	7.4	2.8	6.1	1.9
7	5.0	3.2	1.2	0.2	6.8	2.8	4.8	1.4	7.6	3.0	6.6	2.1
8	4.6	3.2	1.4	0.2	6.1	2.8	4.7	1.2	7.7	2.8	6.7	2.0
9	5.0	3.3	1.4	0.2	4.9	2.4	3.3	1.0	6.2	3.4	5.4	2.3
10	4.8	3.4	1.9	0.2	5.8	2.7	3.9	1.2	7.7	3.0	6.1	2.3
11	4.8	3.0	1.4	0.1	5.8	2.6	4.0	1.2	6.8	3.0	5.5	2.1
12	5.0	3.5	1.3	0.3	5.5	2.4	3.7	1.0	6.4	2.7	5.3	1.9
13	5.1	3.3	1.7	0.5	6.7	3.0	5.0	1.7	5.7	2.5	5.0	2.0
14	5.0	3.4	1.5	0.2	5.7	2.8	4.1	1.3	6.9	3.1	5.1	2.3
15	5.1	3.8	1.9	0.4	6.7	3.1	4.4	1.4	5.9	3.0	5.1	1.8
16	4.9	3.0	1.4	0.2	5.5	2.3	4.0	1.3	6.3	3.4	5.6	2.4
17	5.3	3.7	1.5	0.2	5.1	2.5	3.0	1.1	5.8	2.7	5.1	1.9
18	4.3	3.0	1.1	0.1	6.6	2.9	4.6	1.3	6.3	2.7	4.9	1.8
19	5.5	3.5	1.3	0.2	5.0	2.3	3.3	1.0	6.0	3.0	4.8	1.8
20	4.8	3.4	1.6	0.2	6.9	3.1	4.9	1.5	7.2	3.2	6.0	1.8
21	5.2	3.4	1.4	0.2	5.0	2.0	3.5	1.0	6.2	2.8	4.8	1.8
22	4.8	3.1	1.6	0.2	5.6	3.0	4.5	1.5	6.9	3.1	5.4	2.1
23	4.9	3.6	1.4	0.1	5.6	3.0	4.1	1.3	6.7	3.1	5.6	2.4
24	4.6	3.1	1.5	0.2	5.8	2.7	4.1	1.0	6.4	3.1	5.5	1.8
25	5.7	4.4	1.5	0.4	6.3	2.3	4.4	1.3	5.8	2.7	5.1	1.9
26	5.7	3.8	1.7	0.3	6.1	3.0	4.6	1.4	6.1	3.0	4.9	1.8
27	4.8	3.0	1.4	0.3	5.9	3.0	4.2	1.5	6.0	2.2	5.0	1.5
28	5.2	4.1	1.5	0.1	6.0	2.7	5.1	1.6	6.4	3.2	5.3	2.3
29	4.7	3.2	1.6	0.2	5.6	2.5	3.9	1.1	5.8	2.8	5.1	2.4
30	4.5	2.3	1.3	0.3	6.7	3.1	4.7	1.5	6.9	3.2	5.7	2.3
31	5.4	3.4	1.7	0.2	6.2	2.2	4.5	1.5	6.7	3.0	5.2	2.3
32	5.0	3.0	1.6	0.2	5.9	3.2	4.8	1.8	7.7	2.6	6.9	2.3
33	4.6	3.4	1.4	0.3	6.3	2.5	4.9	1.5	6.3	2.8	5.1	1.5
34	5.4	3.9	1.3	0.4	6.0	2.9	4.5	1.5	6.5	3.0	5.2	2.0
35	5.0	3.6	1.4	0.2	5.6	2.7	4.2	1.3	7.9	3.8	6.4	2.0
36	5.4	3.9	1.7	0.4	6.2	2.9	4.3	1.3	6.1	2.6	5.6	1.4
37	4.6	3.6	1.0	0.2	6.0	3.4	4.5	1.6	6.4	2.8	5.6	2.1
38	5.1	3.8	1.5	0.3	6.5	2.8	4.6	1.5	6.3	2.5	5.0	1.9
39	5.8	4.0	1.2	0.2	5.7	2.8	4.5	1.3	4.9	2.5	4.5	1.7
40	5.4	3.7	1.5	0.2	6.1	2.9	4.7	1.4	6.8	3.2	5.9	2.3
41	5.0	3.4	1.6	0.4	5.5	2.5	4.0	1.3	7.1	3.0	5.9	2.1
42	5.4	3.4	1.5	0.4	5.5	2.6	4.4	1.2	6.7	3.3	5.7	2.5
43	5.1	3.7	1.5	0.4	5.4	3.0	4.5	1.5	6.3	2.9	5.6	1.8
44	4.4	2.9	1.4	0.2	6.3	3.3	4.7	1.6	6.5	3.0	5.5	1.8
45	5.5	4.2	1.4	0.2	5.2	2.7	3.9	1.4	6.5	3.0	5.8	2.2
46	5.1	3.4	1.5	0.2	6.4	2.9	4.3	1.3	7.3	2.9	6.3	1.8
47	4.7	3.2	1.3	0.2	6.6	3.0	4.4	1.4	6.7	2.5	5.8	1.8
48	4.9	3.1	1.5	0.1	5.7	2.6	3.5	1.0	5.6	2.8	4.9	2.0
49	5.2	3.5	1.5	0.2	6.1	2.8	4.0	1.3	6.4	2.8	5.6	2.2
50	5.1	3.5	1.4	0.2	6.0	2.2	4.0	1.0	6.5	3.2	5.1	2.0

Table 1.7: **Body mass**: Height and weight of twenty-two individuals.

Individual	Height, cm	Weight, kg
p1	160	63
p2	160	61
p3	165	64
p4	165	67
p5	164	65
p6	164	62
p7	157	60
p8	158	60
p9	175	75
p10	173	74
p11	180	79
p12	185	84
p13	160	67
p14	160	71
p15	170	73
p16	170	76
p17	180	82
p18	180	85
p19	175	78
p20	174	77
p21	171	75
p22	170	75

The best thing would be if a computer could derive the bmi based decision rule itself, which may not be necessarily the case since the bmi is defined universally whereas only a very limited data set is presented here. One would obviously have to consider whether a linear description could be derived such as the following existing rule of thumb: a man is overwheight if the difference between his height in cm and weight in kg is greater than one hundred. A man 175 cm in height should normally weigh 75 kg or less according to this rule.

Once again it should be pointed out that non-linear transformations supplied by machine learning tools for better prediction may be not necessarily usable for the purposes of description.

The Body mass data set is analyzed on pp. 205, 213, 242.

1.1.3 Association

Revealing associations between different aspects of phenomena is one of the most important goals of classification. Clustering as a classification of empirical data also can do the job. A relation between different aspects of a phenomenon in question can be established if the same clusters are well described twice,

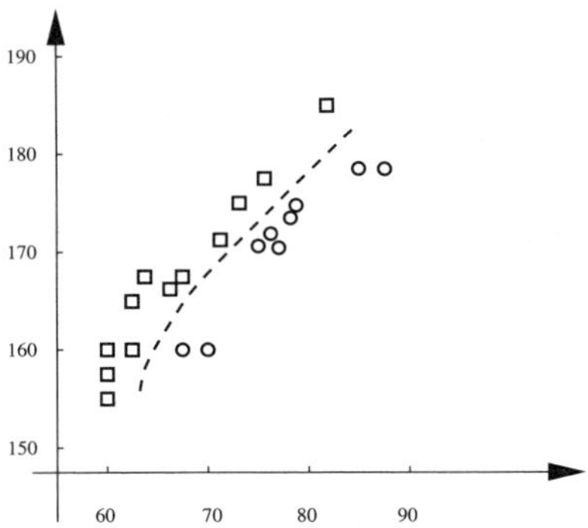

Figure 1.2: Twenty-two individuals at the height-weight plane.

each description related to one of the aspects. Different descriptions of the same cluster are then obviously linked as those referring to the same contents, though possibly with different errors.

Digits and patterns of confusion between them

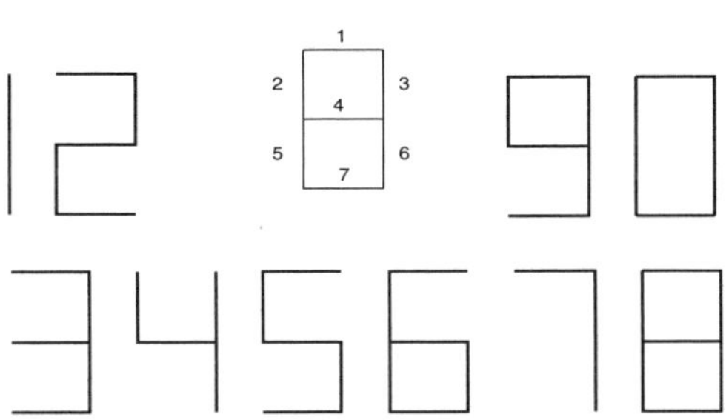

Figure 1.3: Styled digits formed by segments of the rectangle.

The rectangle in the upper part of Figure 1.3 is used to draw numeral digits around it in a styled manner of the kind used in digital electronic devices. Seven binary presence/absence variables e1, e2,..., e7 in Table 1.8 correspond to the numbered segments on the rectangle in Figure 1.3.

Although the digit character images may seem arbitrary, finding patterns of similarity in them may be of interest in training operators dealing with digital numbers.

Table 1.8: **Digits:** Segmented numerals presented with seven binary variables corresponding to presence/absence of the corresponding edge in Figure 1.3.

Digit	e1	e2	e3	e4.	e5	e6	e7
1	0	0	1	0	0	1	0
2	1	0	1	1	1	0	1
3	1	0	1	1	0	1	1
4	0	1	1	1	0	1	0
5	1	1	0	1	0	1	1
6	1	1	0	1	1	1	1
7	1	0	1	0	0	1	0
8	1	1	1	1	1	1	1
9	1	1	1	1	0	1	1
0	1	1	1	0	1	1	1

Table 1.9: **Confusion:** Confusion between the segmented numeral digits.

	Response									
Stimulus	1	2	3	4	5	6	7	8	9	0
1	877	7	7	22	4	15	60	0	4	4
2	14	782	47	4	36	47	14	29	7	18
3	29	29	681	7	18	0	40	29	152	15
4	149	22	4	732	4	11	30	7	41	0
5	14	26	43	14	669	79	7	7	126	14
6	25	14	7	11	97	633	4	155	11	43
7	269	4	21	21	7	0	667	0	4	7
8	11	28	28	18	18	70	11	577	67	172
9	25	29	111	46	82	11	21	82	550	43
0	18	4	7	11	7	18	25	71	21	818

Results of a psychological experiment on confusion between the segmented numerals are in Table 1.9. A digit appeared on a screen for a very short time (stimulus), and an individual was asked to report what was the digit (response). The response frequencies of digits versus shown stimuli stand in the rows of Table 1.9 [90].

The problem is to find general patterns in confusion and to interpret them in terms of the segment presence-absence variables in Digits data Table 1.8. If the found interpretation can be put in a theoretical framework, the patterns can be considered as empirical reflections of theoretically substantiated classes. Patterns of confusion would show the structure of the phenomenon. Interpretation of the clusters in terms of the drawings, if successful, would allow us to see what relation may exist between the patterns of drawing and confusion.

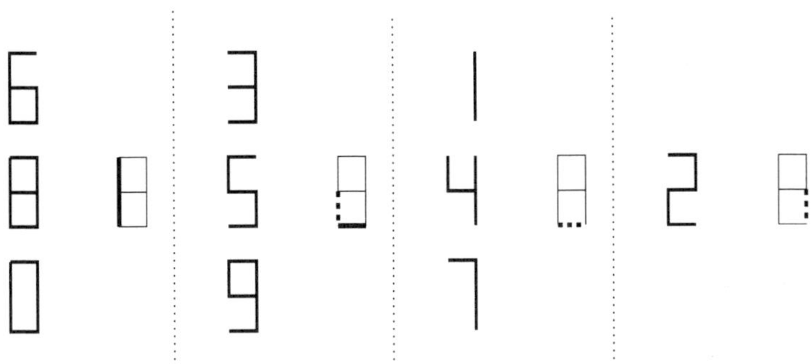

Figure 1.4: Visual representation of four Digits confusion clusters: solid and dotted lines over the rectangle show distinctive features that must be present in or absent from all entities in the cluster.

Indeed, four major confusion clusters can be distinguished in the Digits data, as will be found in section 4.4.2 and described in section 6.3 (see pp. 73, 129, 133 and 134 for computations on these data). On Figure 1.4 these four clusters are presented with distinctive features shown with segments defining the drawing of digits. We can see that all relevant features are concentrated on the left and down the rectangle. It remains to be seen if there is any physio-psychological mechanism behind this and how it can be utilized.

Moreover, it appears the attributes in Table 1.8 are quite relevant on their own, pinpointing the same patterns that have been identified as those of confusion. This can be clearly seen in Figure 1.5, which illustrates a classification tree for Digits found using an algorithm for conceptual clustering presented in section 4.3. On this tree, clusters are the terminal boxes and interior nodes are labeled by the features involved in classification. The coincidence of the drawing clusters with confusion patterns indicates that the confusion is caused by the segment features participating in the tree. These appear to be the same features in both Figure 1.4 and Figure 1.5.

Literary masterpieces

The data in Table 1.10 reflect the language and style features of eight novels by three great writers of the nineteenth century. Two language features are:

1) LenSent - Average length of (number of words in) sentences;

2) LenDial - Average length of (number of sentences in) dialogues. (It is assumed that longer dialogues are needed if the author uses dialogue as a device to convey information or ideas to the reader.)

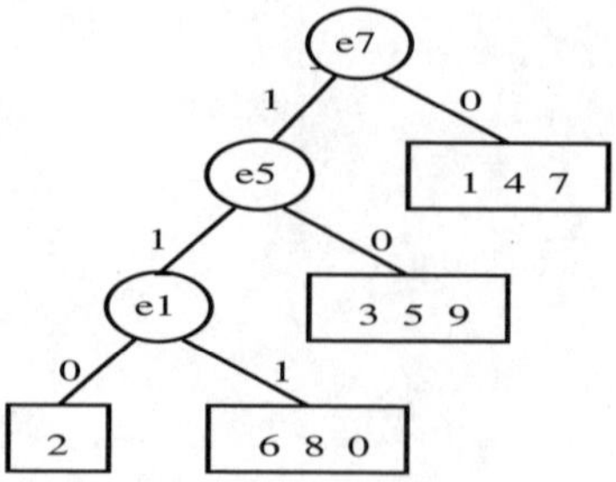

Figure 1.5: The conceptual tree of Digits.

Table 1.10: **Masterpieces:** Masterpieces of 19th century: the first three by Charles Dickens (1812–1870), the next three by Mark Twain (1835–1910), and the last two by Leo Tolstoy (1828–1910).

Title	LenSent	LenDial	NChar	SCon	Narrative
Oliver Twist	19.0	43.7	2	No	Objective
Dombey and Son	29.4	36.0	3	No	Objective
Great Expectations	23.9	38.0	3	No	Personal
Tom Sawyer	18.4	27.9	2	Yes	Objective
Huckleberry Finn	25.7	22.3	3	Yes	Personal
Yankee at King Arthur	12.1	16.9	2	Yes	Personal
War and Peace	23.9	30.2	4	Yes	Direct
Anna Karenina	27.2	58.0	5	Yes	Direct

Features of style:

3) NChar - Number of principal characters (the larger the number the more themes raised);

4) SCon - Yes or No depending on the usage of the stream of conscience techniques;

5) Narrative - The narrative style is a qualitative feature categorized as: (a) Personal (if the narrative comes from the mouth of a character such as Pip in "Great Expectations" by Charles Dickens), or (b) Objective (if the subject develops mainly through the behavior of the characters and other indirect means), or (c) Direct (if the author prefers to directly intervene with the comments and explanations).

As we have seen already with the Digits data, features are not necessarily quantitative. They also can be categorical, such as SCon, a binary variable, or Narrative, a nominal variable.

The data in Table 1.10 can be utilized to advance two of the clustering goals:

1. **Structurization:** To cluster the set of masterpieces and intensionally describe clusters in terms of the features. We expect the clusters to accord to the three authors and convey features of their style.

2. **Association:** To analyze interrelations between two aspects of prose writing: (a) linguistic (presented by LenSent and LenD), and (b) the author's narrative style (the other three variables). For instance, we may find clusters in the linguistic features space and conceptually describe them in terms of the narrative style features. The number of entities that do not satisfy the description will score the extent of correlation. We expect, in this particular case, to have a high correlation between these aspects, since both must depend on the same cause (the author) which is absent from the feature list (see page 104).

This data set is used for illustration of many concepts and methods described further on; see pp. 61, 62, 78, 79, 80, 81, 84, 89, 104, 105, 162, 182, 193, 195, 197.

1.1.4 Generalization

Generalization, or overview, of data is a (set of) statement(s) about properties of the phenomenon reflected in the data under consideration. To make a generalization with clustering, one may need to do a multistage analysis: at first, structure the entity set; second, describe clusters; third, find associations between different aspects.

Probably one of the most exciting applications of this type can be found in the newly emerging area of text mining [139]. With the abundance of text information flooding every Internet user, the discipline of text mining is flourishing. A traditional paradigm in text mining is underpinned by the concept of the key word. The key word is a string of symbols (typically corresponding to a language word or phrase) that is considered important for the analysis of a pre-specified collection of texts. Thus, first comes a collection of texts defined by a meaningful query such as "recent mergers among insurance companies" or "medieval Britain." (Keywords can be produced by human experts in the domain or from statistical analyses of the collection.) Then a virtual or real text-to-keyword table can be created with keywords treated as features. Each of the texts (entities) can be represented by the number of occurrences of each of the keywords. Clustering of such a table may lead to finding subsets of texts covering different aspects of the subject.

This approach is being pursued by a number of research and industrial groups, some of which have built clustering engines on top of Internet search engines: given a query, such a clustering engine singles out several dozen of the most relevant web pages, resulting from a search by a search engine such as

Table 1.11: List of eleven features I-XI and their categories with respect to five aspects of a Bribery situation.

Actor	Service	Interaction	Environment
I. Office	**III. Type of service**	**V. Initiator**	**IX. Condition**
1. Enterprise	1. Obstr. of justice	1. Client	1. Regular routine
2. City	2. Favors	2. Official	2. Monitoring
3. Region	3. Extortion		3. Sloppy regulations
4. Federal	4. Category change		4. Irregular
	5. Cover-up		
II. Client	**IV. Occurrence**	**VI. Bribe level**	**X. Branch**
1. Individual	1. Once	1. \$10K or less	1. Government
2. Business	2. Multiple	2. Up to \$100K	2. Law enforcement
		3. $\geq$\$100K	3. Other
		VII. Type	**XI. Punishment**
		1. Infringement	1. None
		2. Extortion	2. Administrative
			3. Arrest
		VIII. Network	4. Arrest followed by release
		1. None	5. Arrest with imprisonment
		2. Within office	
		3. Between offices	
		4. Clients	

Google or Yahoo, finds keywords or phrases in the corresponding texts, clusters web pages according to the keywords used as features, and then describes clusters in terms of the most relevant keywords or phrases. Two top web sites which have been found from searching for "clustering engines" with Google on 29 June 2004 in London are Vivisimo at hhtp://vivisimo.com and iBoogie at http://iboogie.tv. The former is built on top of ten popular search engines and can be used for partitioning web pages from several different sources such as "Web" or "Top stories," the latter maintains several dozen languages and presents a hierarchical classification of selected web pages. In response to the query "clustering" Vivisimo produced 232 web pages in a "Web" category and 117 in a "Top news" category. Among top news the most populated clusters were "Linux" (16 items), "Stars" (12), and "Bombs" (11). Among general web sites the most numerous were "Linux" (25), "Search, Engine" (21), "Computing" (22), etc. More or less random web sites devoted to individual papers or scientists or scientific centers or commercial companies have been listed under categories "Visualization" (12), "Methods" (7), "Clustering" (7), etc. Such categories as "White papers" contained pages devoted to both computing clusters and cluster analysis. Similar results, though somewhat more favourable towards clustering as data mining, have been produced with iBoogie. Its cluster "Cluster" (51) was further divided into categories such as "computer" (10) and "analysis" (5). Such categories as "software for clustering" and "data clustering" have been presented too to refer to a random mix of 24 and 20 web sites respectively.

The activity of generalization so far mainly relies on human experts who supply understanding of a substantive area behind the text corpus. Human experts develop a text-to-feature data table that can be further utilized for generalization. Such is a collection of 55 articles on Bribery cases from central Russian newspapers 1999-2000 presented in Table 1.12 according to [97]. The features reflect the following fivefold structure of bribery situations: two interacting sides - the office and the client, their interaction, the corrupt service rendered, and the environment in which it all occurs.

These structural aspects can be characterized by eleven features that can be recovered from the newspaper articles; they are presented in Table 1.11.

To show how these features can be applied to a newspaper article, let us quote an article that appeared in a newspaper called "Kommersant" on 20 March 1999 (translated from Russian):

Mayor of a coal town under arrest
Thursday this week, Mr Evgeny Parshukov, Mayor of town Belovo near Kemerovo, was arrested under a warrant issued by the region attorney, Mr. Valentin Simuchenkov. The mayor is accused of receiving a bribe and abusing his powers for wrongdoing. Before having been elected to the mayoral post in June 1997, he received a credit of 62,000 roubles from Belovo Division of KUZBASS Transport Bank to support his election campaign. The Bank then cleared up both the money and interest on it, allegedly because after his election Mr. Parshukov ordered the Finance Department of the town administration, as well as all municipal organisations in Belovo, to move their accounts into the Transport Bank. Also, the attorney office claims that in 1998 Mr. Parshukov misspent 700,000 roubles from the town budget. The money came from the Ministry of Energy specifically aimed at creating new jobs for mine workers made redundant because their mines were getting closed. However, Mr. Parshukov ordered to lend the money at a high interest rate to the Municipal Transport agency. Mr. Parshukov doesn't deny the facts. He claims however that his actions involve no crime.

A possible coding of the eleven features in this case constitutes the contents of row 29 in Table 1.12. The table presents 55 cases that could be more or less unambiguously coded (from the original 66 cases [98]).

The prime problem here is similar to those in the Market towns and Digits data: to see if there are any patterns at all. To generalize, one has to make sense of patterns in terms of the features. In other words, we are interested in getting a synoptic description of the data in terms of clusters which are to be found and described.

On the first glance, no structure exists in the data. Nor could the scientists

Table 1.12: **Bribery:** data with features from Table 1.11.

Case	Of	Cl	Serv	Occ	Ini	Br	Typ	Net	Con	Branch	Pun
1	2	2	2	2	1	2	1	3	3	1	5
2	2	1	5	2	1	1	1	3	2	1	5
3	2	2	2	1	1	3	1	1	3	1	4
4	1	2	3	1	2	1	2	1	3	1	3
5	1	1	4	2	2	1	1	4	3	3	3
6	1	2	3	1	2	2	2	1	3	1	5
7	3	2	1	1	2	3	2	3	2	2	5
8	2	2	4	2	1	1	1	2	1	2	5
9	1	2	3	1	2	1	2	1	1	1	5
10	1	2	3	1	2	1	2	1	1	1	5
11	2	2	5	1	2	3	2	2	2	1	5
12	2	2	1	1	2	2	1	1	4	2	5
13	3	2	1	1	1	3	1	1	4	2	2
14	2	1	4	1	2	1	1	2	1	2	5
15	3	2	2	1	2	1	1	2	3	1	5
16	2	1	4	2	2	1	1	1	1	3	3
17	4	2	2	1	1	2	1	1	4	1	5
18	2	2	5	1	1	2	2	2	3	2	5
19	2	2	5	1	2	1	2	1	3	2	5
20	2	2	1	1	1	2	1	4	3	2	5
21	1	2	3	1	2	2	2	1	3	1	5
22	1	2	2	2	2	2	1	1	4	1	3
23	1	1	4	1	2	1	1	1	1	3	5
24	3	2	5	1	2	1	1	2	2	2	2
25	2	1	2	2	2	1	2	1	3	3	5
26	1	2	5	1	2	2	1	1	3	1	3
27	1	1	4	2	1	2	1	2	4	3	5
28	1	1	5	1	2	1	1	1	2	3	5
29	2	2	2	1	2	2	1	1	3	1	5
30	2	2	3	1	2	2	2	3	3	1	5
31	2	2	3	1	2	1	2	3	3	1	5
32	4	2	2	1	2	1	1	1	3	1	5
33	3	1	1	1	2	1	1	1	4	2	3
34	3	2	1	2	1	1	1	2	4	2	5
35	3	1	3	2	2	1	2	3	4	2	3
36	2	1	3	1	2	2	1	3	1	2	5
37	2	2	5	1	2	3	1	1	2	2	5
38	2	1	1	1	2	2	1	3	4	2	4
39	2	2	1	1	1	1	1	4	4	2	5
40	1	2	3	2	2	1	2	1	3	2	5
41	1	1	4	2	2	1	2	1	2	3	5
42	1	1	4	2	2	1	1	1	3	3	5
43	2	1	1	1	1	1	1	4	4	2	5
44	3	2	5	2	2	1	1	1	2	2	5
45	2	1	1	1	1	1	1	2	4	2	5
46	3	2	5	2	2	1	1	2	2	2	5
47	2	1	1	1	2	1	2	2	4	2	1
48	2	1	1	1	2	1	1	1	4	2	5
49	3	1	2	1	1	2	1	3	4	1	5
50	3	1	1	1	2	1	1	1	4	2	5
51	2	1	2	1	2	1	1	1	4	2	5
52	3	2	1	1	2	1	2	1	2	2	5
53	2	2	5	1	2	2	2	2	2	2	5
54	2	2	5	2	2	2	1	2	2	2	5
55	2	2	5	1	2	3	2	1	2	2	2

specializing in the research of corruption see any. However, after applying an intelligent version of the algorithm K-Means as described later in example 3.20, section 3.3, a rather simple core structure could be found that is defined by just two features and determines all other aspects. The results provide for a really short generalization: "It is the branch of government that determines which of the five types of corrupt services are rendered: Local government → Favors or Extortion; Law enforcement → Obstruction of Justice or Cover-Up; and Other → Category Change." A detailed discussion is given in examples on pp. 95, 106 and 147.

1.1.5 Visualization of data structure

Visualization is considered a rather vague area involving psychology, cognitive sciences and other disciplines, which is rapidly developing. In the current thinking, the subject of data visualization is defined as creation of mental images to gain insight and understanding [125]. This, however, seems too wide and includes too many non-operational images such as realistic and surrealistic paintings. In our presentation, we take on a more operational view and consider that data visualization is an activity related to mapping data onto a known ground image such as a coordinate plane, geography map, or a genealogy tree in such a way that properties of the data are reflected in the structure of the ground image.

Among ground images, the following are the most popular: geographical maps, networks, 2D displays of one-dimensional objects such as graphs or pie-charts or histograms, 2D displays of two-dimensional objects, and block structures. Sometimes, the very nature of the data suggests what ground image should be used. All of these can be used with clustering, and we are going to review most of them except for geographical maps.

One-dimensional data

One-dimensional data over pre-specified groups or found clusters can be of two types: (a) the distribution of entities over groups and (b) values of a feature within clusters. Accordingly, there can be two types of visual support for these.

Consider, for instance, groups of the Market town data defined by the population. According to Table 1.1 the population ranges approximately between 2000 and 24000 habitants. Let us divide the range in five equal intervals, bins, that are defined thus to have size (24000-2000)/5=4400 and bound points 6400, 10800, 15200, and 19600.

In Table 1.13 the data of the groups are displayed: their absolute and relative sizes and also the average numbers of Banks and the standard deviations within them. For the definitions of the average and standard deviation see section 2.1.2.

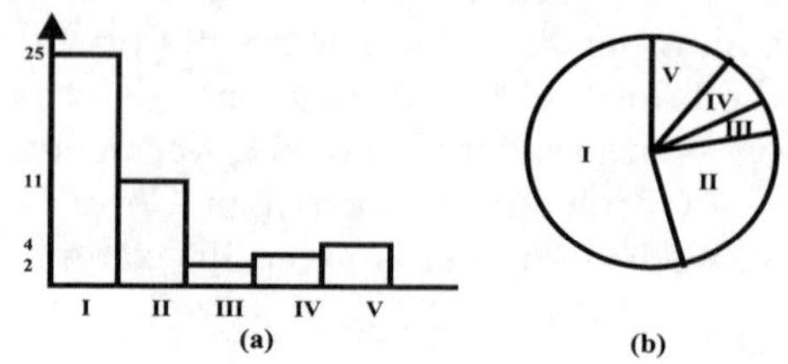

Figure 1.6: Histogram (a) and pie-chart (b) presenting the distribution of Population over five equally sized bins in Market data.

Figure 1.6 shows two traditional displays for the distribution: a *histogram* (part (a) on the left) in which bars are proportional to the group sizes and a *pie-chart* in which a pie is partitioned into slices proportional to the cluster sizes (part (b) on the right). These two point to different features of the distribution. The histogram positions the categories along the horizontal axis, thus providing for a possibility to see the distribution's shape, which can be quite useful when the categories have been created as interval bins of a quantitative feature, as is this case. The pie-chart points to the fact that the group sizes sum up to the total so that one can see what portions of the pie account for different categories.

One-dimensional data within groups

To visualize a quantitative feature within pre-specified groups, *box-plots* and *stick-plots* are utilized. They show within-cluster central values and their dis-

Table 1.13: **Population groups:** Data of the distribution of population groups and numbers of banks within them.

Group	Size	Frequency, %	Banks	Std Banks
I	25	55.6	1.80	1.62
II	11	24.4	4.82	2.48
III	2	4.4	5.00	1.00
IV	3	6.7	12.00	5.72
V	4	8.9	12.50	1.12
Total	45	100	4.31	4.35

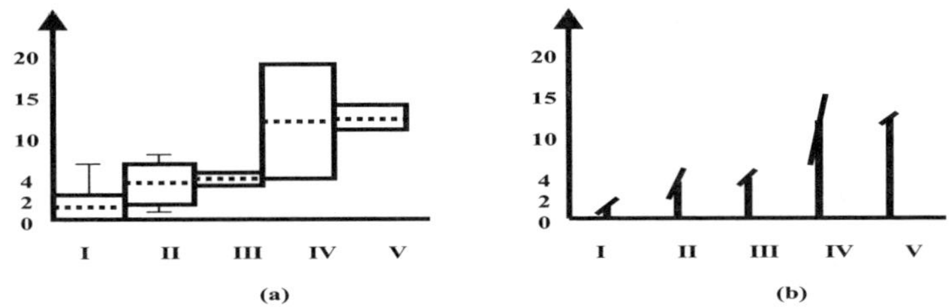

Figure 1.7: Box-plot (a) and stick-plot (b) presenting the feature Bank over the five bins in Market data.

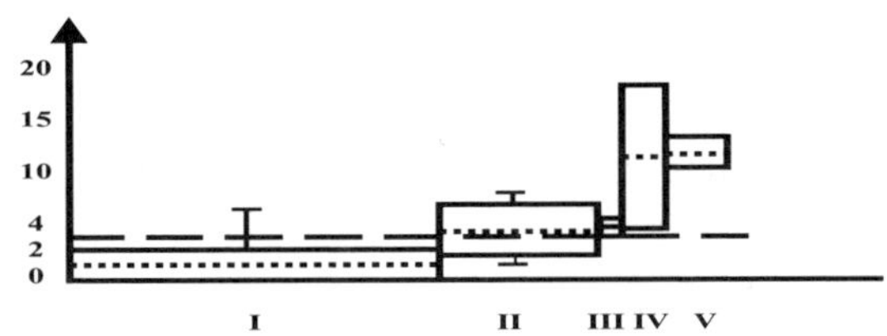

Figure 1.8: Box-plot presenting the feature Bank over five bins in Market data along with bin sizes.

persion, which can be done in different ways. Figure 1.7 presents a box-plot (a) and stick-plot (b) of feature Bank within the five groups defined above as Population bins. The box-plot on Figure 1.7 (a) represents each group as a box bounded by its 10% percentile values separating extreme 10% cases both on the top and bottom of the feature Bank range. The real within group ranges are shown by "whiskers" that can be seen above and below the boxes of groups I and II; the other groups have no whiskers because of too few entities in each of them. A line within each box shows the within-group average. The stick-plot on Figure 1.7 (b) represents the within-group averages by "sticks," with their "whiskers" proportional to the standard deviations.

Since the displays are in fact two-dimensional, both features and distributions can be shown on a box-plot simultaneously. Figure 1.8 presents a box-plot of the feature Bank over the five bins with the box widths made proportional to the group sizes. This time, the grand mean is also shown by the horizontal dashed line.

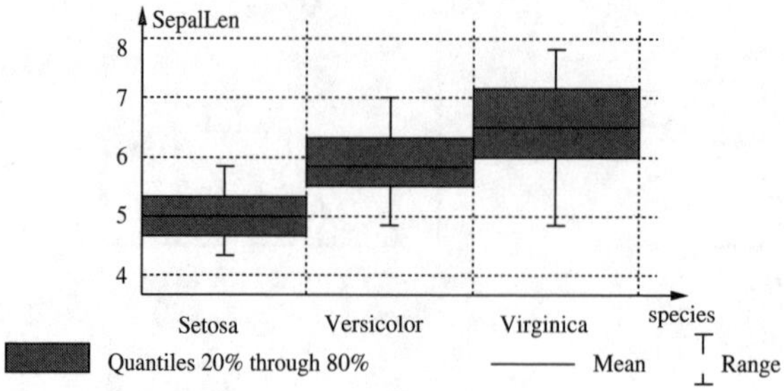

Figure 1.9: Box-plot of three classes of Iris specimens from Table 1.6 over the sepal length w1; the classes are presented by both the percentile boxes and within cluster range whiskers; the choice of percentiles can be adjusted by the user.

A similar box-plot for the three genera in the Iris data is presented in Figure 1.9. This time the percentiles are taken at 20%.

Two-dimensional display

A traditional two-dimensional display of this type is the so-called *scatter-plot*, representing all the entity points in a plane generated by two of the variables or linear combinations of the variables such as principal components (for a definition of principal components see section 5.1.3). A scatter-plot at the plane of two variables can be seen in Figure 1.2 for the Body mass data on page 13. A scatter-plot in the space of two first principal components is presented in Figure 1.10: the Iris specimens are labelled by the class number (1, 2, or 3); centroids are gray circles; the most deviate entities (30 in class 1, 32 in class 2, and 39 in class 3) are shown in boxes. For an explanation of the principal components see section 5.1.3. The scatter-plot illustrates that two of the classes are somewhat interwoven.

Block-structure

A block-structure is a representation of the data table as organized in larger blocks of a specific pattern with transpositions of rows and/or columns. In principle one can imagine various block patterns [47], of which the most common is a pattern formed by the largest entry values.

Figure 1.11 presents an illustrative example (described in [125]). In part A, results of seven treatments (denoted by letters from a to g) applied to each of ten crops denoted by numerals are presented: gray represents a success and blank space failure. The pattern of gray seems rather chaotic in table A. However, it becomes very much orderly when appropriate rearrangements of rows and

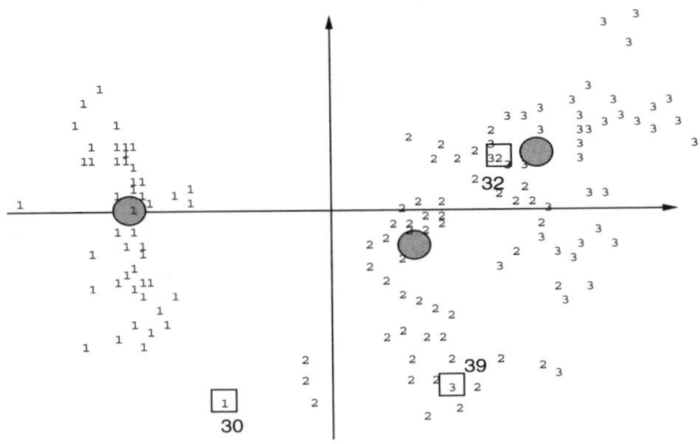

Figure 1.10: Scatter-plot of Iris specimens in the plane of the first two principal components.

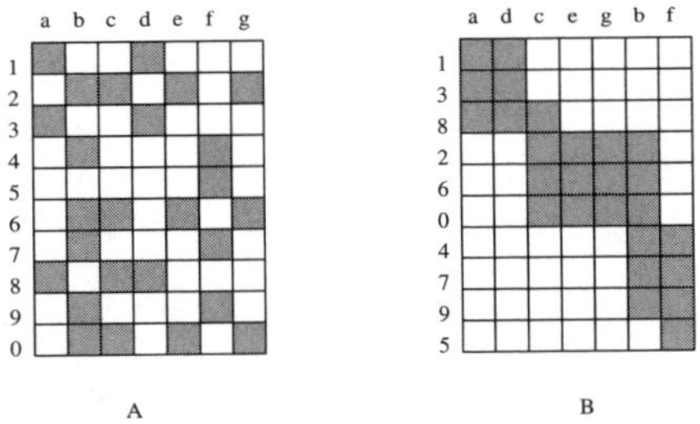

Figure 1.11: Visibility of the matrix block structure with a rearrangement of rows and columns.

columns are performed. Part B of the Figure clearly demonstrates a visible block structure in the matrix, that can be interpreted as mapping specific sets of treatments to different sets of crops, which can be exploited, for instance, in specifying adjacent locations for crops.

Visualization of block structures by reordering rows and/or columns is popular in the analysis of gene expression data [20] and ecology [53].

A somewhat more realistic example is shown in Figure 1.12 (a) representing a matrix of value transferred between nine industries during a year: the (i,j)-th entry is gray if the transfer from industry i to industry j is greater than a specified threshold and blank otherwise. Figure 1.12 (b) shows a block structure pattern that becomes visible when the order of industries from 1 to 9 changes for the order 1-6-9-4-2-5-8-3-7, which is achieved with the reordering of both rows and columns of the matrix. The reordering is made simultaneously on both rows and columns because both represent the same industries, both as sources (rows) and targets (columns) of the value transfer. We can discern four blocks of different patterns (1-6-9, 4, 2-5-8, 3-7) in Figure 1.12 (b). The structure of the transfers between the blocks can be captured in a graph presented in Figure 1.12 (c).

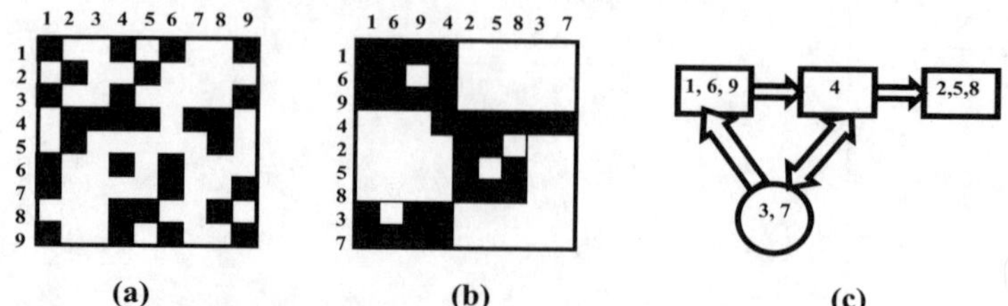

Figure 1.12: Value transfer matrix presented with only entries greater than a threshold (a); the same matrix, with rows and columns simultaneously reordered, is in (b); in (c), the structure is represented as a graph.

Structure

A simple structure such as a chain or a tree or just a small graph, whose vertices (nodes) correspond to clusters and edges to associations between them, is a frequent tool in data visualization.

Two examples are presented in Figure 1.13: a tree structure over clusters reflecting common origin is shown in part (a) and a graph corresponding to the block structure of Figure 1.12 (c) is shown in part (b) to reflect links between clusters of industries in the production process.

A similar tree structure is presented on Figure 1.5 on page 16 illustrating a classification tree for Digits. Tree leaves, the terminal boxes, show clusters as entity sets; the features are shown along corresponding branches; the entire structure illustrates the relation between clusters in such a way that any combination of the segments can be immediately identified and placed into a corresponding cluster or not identified at all if it is not shown on the tree.

Visualization using an inherent topology

In many cases the entities come from an image themselves – such as in the cases of analysis of satellite images or topographic objects. For example, consider the Digit data set: all the integer symbols are associated with segments of the generating rectangle on Figure 1.3, page 13. Clusters of such entities can be visualized with the generating image.

Figure 1.4 visualizes clusters of digits along with their defining features resulting from analyses conducted later in example 4.43 (page 134) as parts of the generating rectangle. There are four major confusion clusters in the

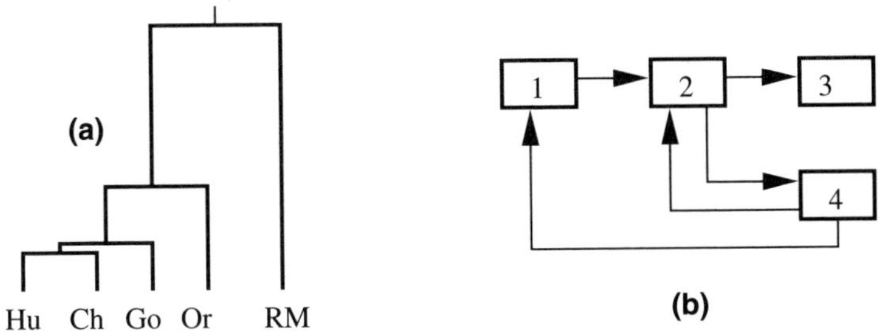

Figure 1.13: Visual representation of relations between clusters: (a) the evolutionary structure of the Primates genera according to distances in Table 1.2; (b) interrelation between clusters of industries according to Figure 1.12 (c).

Digits data of Figure 1.3 that are presented with distinctive features shown with segments defining the drawing of digits.

1.2 Bird's-eye view

This section contains general remarks on clustering and can be skipped on the first reading.

1.2.1 Definition: data and cluster structure

After looking through the series of exemplary problems in the previous section, we can give a more formal definition of clustering than that in the Preface: Clustering is a discipline devoted to revealing and describing cluster structures in data sets.

To animate this definition, one needs to specify the four concepts involved:

(a) **data**,

(b) **cluster structure**,

(c) **revealing a cluster structure**,

(d) **describing a cluster structure**.

Data

The concept of data refers to any recorded and stored information such as satellite images or time series of prices of certain stocks or survey questionnaires filled in by respondents. Two types of information are associated with data: the data entries themselves, e.g., recorded prices or answers to questions, and meta-data, that is, legends to rows and columns giving meaning to entries. The aspect of developing and maintaining databases of records, taking into account the relations stored in metadata, is very important for data mining [23, 29, 44].

In this text, for the sake of linearity of presentation, we concentrate on a generic data format only, the so-called entity-to-variable table whose entries represent values of pre-specified variables at pre-specified entities.

The variables are synonymously called attributes, features, characteristics, characters and parameters. Such words as case, object, observation, instance, record are in use as synonyms to the term, entity, accepted here.

The data table format of data often arises directly from experiments or observations, from surveys, and from industrial or governmental statistics. This also is a conventional form for presenting database records. Other data types such as signals or images can be modelled in this format, too, via digitalized representation. However, a digital representation, typically, involves much more information than can be kept in a data table format. Especially important is the spatial arrangement of pixels, which is not, typically, maintained in the concept of data table. The contents of a data table are assumed to be invariant under permutations of rows and columns and corresponding metadata.

Another data type traditionally considered in clustering is the similarity or dissimilarity between entities (or features). The concept of similarity is most important in clustering: similar objects are to be put into the same cluster and dissimilar into different clusters. There have been invented dozens of (dis)similarity indices. Some of them nicely fit into theoretical frameworks and will be considered further in the text.

One more data type considered in this text is co-occurrence or flow tables that represent the same substance distributed between different categories such as Confusion data in Table 1.9 in which the substance is the scores of individuals. An important property of this type of data is that any part of the data table, referred to a subset of rows and /or a subset of columns, can be meaningfully aggregated by summing the part of the total flow within the subset of rows (and/or the subset of columns). The sums represent the total flow to, from, and within the subset(s). Thus, problems of clustering and aggregating are naturally linked here. Until recently, this type of data appeared as a result of data analysis rather than input to it. Currently it has become one of the major data formats. Examples are: distributions of households purchasing various commodities or services across postal districts or other regional units, counts of telephone calls across areas, and counts of visits in various categories of web-sites.

Cluster structure

The concept of **cluster** typically refers to a set of entities that is cohesive in such a way that entities within are more similar to each other than to the outer entities.

Three major types of cluster structures are: (a) a single cluster considered against the rest or whole of the data, (b) a partition of the entity set in a set of clusters, and (c) a (nested) hierarchy of clusters.

Of these three, partition is the most conventional, probably because it is relevant to both science and management, the major forces behind scientific developments. A scientist, as well as a manager, wants unequivocal control over the entire universe under consideration. This is why they may wish to partition the entity set into a set of nonoverlapping clusters.

In some situations there is no need for total clustering. The user may be quite satisfied with getting just a single (or few) cluster(s) and leaving the rest completely unclustered. Examples:

(1) a bank manager wants to learn how to discern potential fraudsters from other clients or

(2) a marketing researcher separates a segment of customers prone to purchase a particular product or

(3) a bioinformatician seeks a set of proteins homologous to a query protein sequence.

Incomplete clustering is a recently recognized addition to the body of clustering approaches, very suitable not only at the situations above but also as a tool for conventional partitioning via cluster-by-cluster procedures such as those described in section 5.5.

The hierarchy is the oldest and probably least understood of the cluster structures. To see how important it is, it should suffice to recall that the Aristotelian approach to classification encapsulated in library classifications and biological taxonomies is always based on hierarchies. Moreover, hierarchy underlies most advanced data processing tools such as wavelets and quadtrees. It is ironic then that as a cluster structure in its own right, the concept of hierarchy rarely features in clustering, especially when clustering is confined to the cohesive partitioning of geometric points.

1.2.2 Criteria for revealing a cluster structure

To **reveal a cluster structure** in a data table means to find such clusters that allow the individual characteristics of entities to be substituted by aggregate characteristics of clusters. A cluster structure is revealed by a *method* according to a *criterion* of how well the data are represented by clusters. Criteria and methods are, to an extent, independent from each other so that the same method such as agglomeration or splitting can be used with different criteria.

Criteria usually are formulated in terms of (dis)similarity between entities. This helps in formalizing the major idea that entities within clusters should be similar to each other and those between clusters dissimilar. Dozens of similarity based criteria developed so far can be categorized in three broad classes:

(1) Definition-based,
(2) Index-based, and
(3) Computation-based.

The first category comprises methods for finding clusters according to an explicit definition of a cluster. An example: A cluster is a subset S of entities such that for all i, j in S the similarity between i and j is greater than the similarities between these and any k outside S. Such a property must hold for all entities with no exceptions, which means that well isolated clusters are rather rare in real world data. However, when the definition of cluster is relaxed to include less isolated clusters, too many may then appear. This is why definition-based methods are not popular in practical clustering.

A criterion in the next category involves an index, that is, a numerical function that scores different cluster structures and, in this way, may guide the process of choosing the best. However, not all indices are suitable for obtaining reasonable clusters. Those derived from certain model-based considerations tend to be computationally hard to optimize. Optimizing methods are thus bound to be local and, therefore, heavily reliant on the initial settings, which involve, in the case of K-Means clustering, pre-specifying the number of clusters and the location of their central points. Accordingly, the found cluster structure may be rather far from the global optimum and, thus, must be *validated*. Cluster validation may be done according to internal criteria such as that involved in the optimization process or external criteria comparing the clusters found with those known from external considerations or according to its stability with respect to randomly resampling entities/features. These will be outlined in section 7.5 and exemplified in section 3.3.2.

The third category comprises computation methods involving various heuristics for individual entities to be added to or removed from clusters, for merging or splitting clusters, and so on. Since operations of this type are necessarily local, they resemble local search optimization algorithms, though, typically, have no unique guiding scoring index to follow, thus, can include various tricks making them flexible. However, such flexibility is associated with an increase in the number of ad hoc parameters such as various similarity thresholds and, in this way, turning clustering from a reproducible activity into a kind of magic. Validation of a cluster structure found with a heuristic-based algorithm becomes a necessity.

In this book, we adhere to an index-based principle, which scores a cluster structure against the data from which it has been built. The cluster structure here is used as a device for reconstructing the original data table; the closer the reconstructed data are to the original ones, the better the structure. It is this principle that is called the *data recovery approach* in this book. Many index-based and computation-based clustering methods can be reinterpreted according to the principle, which allows us to see interrelations between different methods and concepts for revealing and analyzing clustering structures. New methods can be derived from the principle too (see Chapter 5, especially sections 5.4-5.6). It should be noted, though, that we will use only the most straightforward rules for reconstructing the data from cluster structures.

1.2.3 Three types of cluster description

Cluster descriptions help in understanding, explaining and predicting clusters. These may come in different formats of which the most popular are the following three: (a) Representative, (b) Tendency, (c) Conceptual description.

A *representative*, or a *prototype*, is an object such as a literary character or a sort of wine or mineral, representing the most typical features of a cluster. This format is useful in giving a meaning to entities that are easily available empirically but difficult to conceptually describe. There is evidence that some aggregate language constructs, such as "fruit," are mentally maintained via prototypes, such as "apple" [74]. In clustering, the representative is usually the most central entity in a cluster.

A *tendency* expresses a cluster's most likely features such as its way of behavior or pattern. It is usually related to the center of gravity of the cluster and its differences from the average. In this respect, the tendency models the concept of type in classification studies.

A *conceptual description* may come in the form of a classification tree built for predicting a class or partition. Another form of conceptual description is an *association*, or *production, rule*, stating that if an object belongs to a cluster then it must have such and such features. Or, vice versa, if an object satisfies the premise, then it belongs in the cluster. The simplest conceptual description of a cluster is a statement of the form "the cluster is characterized by the feature A being between values a1 and a2." The existence of a feature A, which alone is sufficient to distinctively describe a cluster is a rare occurrence of luck in data mining. Typically, features in data are rather superficial and do not express essential properties of entities and thus cannot be the basis of straightforward descriptions.

The subject of cluster description overlaps that of supervised machine learning and pattern recognition. Indeed, given a cluster, having its description may allow one to predict, for new objects, whether they belong to the cluster or not, depending on how much they satisfy the description. On the other hand, a decision rule obtained with a machine learning procedure, especially, for example, a classsification tree, can be considered a cluster description usable for the interpretation purposes. Still the goals are different: interpretation in clustering and prediction in machine learning. However, cluster description is as important in clustering as cluster finding.

1.2.4 Stages of a clustering application

Typically, clustering as a data mining activity involves the following five stages:

A. Developing a data set.

B. Data pre-processing and standardizing.

C. Finding clusters in data.

D. Interpretation of clusters.

E. Drawing conclusions.

To develop a data set one needs to define a substantive problem or issue, however vague it may be, and then determine what data set related to the issue can be collected from an existing database or set of experiments or survey, etc.

Data pre-processing is the stage of preparing data processing by a clustering algorithm; typically, it includes developing a uniform data set, frequently called a 'flat' file, from a database, checking for missing and unreliable entries, rescaling and standardizing variables, deriving a unified similarity measure, etc.

The cluster finding stage involves application of a clustering algorithm and results in a (series of) cluster structure(s) to be presented, along with interpretation aids, to substantive specialists for an expert judgement and interpretation in terms of features, both those utilized for clustering (internal features) and those not utilized (external features). At this stage, the expert may see no relevance in the results and suggest a modification of the data by adding/removing features and/or entities. The modified data is subject to the same processing procedure. The final stage is the drawing of conclusions, with respect to the issue in question, from the interpretation of the results. The more focussed are the regularities implied by the findings, the better the quality of conclusions.

There is a commonly held opinion among specialists in data analysis that the discipline of clustering concerns only the proper clustering stage C while the other four are the concern of specialists in the substance of the particular issue for which clustering is performed. Indeed, typically, clustering results can not and are not supposed to solve the entire substantive problem, but rather relate to an aspect of it.

On the other hand, clustering algorithms are supposedly most applicable to situations and issues in which the user's knowledge of the domain is more superficial than profound. What are the choices regarding data pre-processing, initial settings in clustering and interpretation of results – facing the laymen user who has an embryonic knowledge of the domain? More studies and experiments? In most cases, this is not practical advice. Sometimes a more viable strategy would be to better utilize properties of the clustering methods at hand.

At this stage, no model-based recommendations can be made about the initial and final stages, A and E. However, the data recovery approach does allow us to use the same formalisms for tackling not stage C only, but also B and D; see sections 2.4, 4.3 and 6.3 for related prescriptions and discussions.

1.2.5 Clustering and other disciplines

The concepts involved make clustering a multidisciplinary activity on its own, regardless of its many applications. In particular,

1. **Data** relates to database, data structure, measurement, similarity and dissimilarity, statistics, matrix theory, metric and linear spaces, graphs, data analysis, data mining, etc.

2. **Cluster structure** relates to discrete mathematics, abstract algebra, cognitive science, graph theory, etc.

3. **Revealing** cluster structures relates to algorithms, matrix analysis, optimization, computational geometry, etc.

4. **Describing** clusters relates to machine learning, pattern recognition, mathematical logic, knowledge discovery, etc.

1.2.6 Different perspectives of clustering

Clustering is a discipline on the intersection of different fields and can be viewed from different angles, which may be sometimes confusing because different perspectives may contradict each other. A question such as, "How many clusters are out there?," which is legitimate in one perspective, can be meaningless in the other. Similarly, the issue of validation of clusters may have different solutions in different frameworks. The author finds it useful to distinguish between the perspectives supplied by statistics, machine learning, data mining and classification.

Statistics perspective

Statistics tends to view any data table as a sample from a probability distribution whose properties or parameters are to be estimated with the data. In the case of clustering, clusters are supposed to be associated with different probabilistic distributions which are intermixed in the data and should be recovered from it.

Within this approach, such questions as "How many clusters are out there?" and "How to preprocess the data?" are well substantiated and can be dealt with according to the assumptions of the underlying model.

In many cases the statistical paradigm suits quite well and should be applied as the one corresponding most to what is called the scientific method: make a hypothesis of the phenomenon in question, then look for relevant data and check how the hypothesis fits them.

A trouble with this approach is that in most cases clustering is applied to phenomena of which almost nothing is known, not only of their underlying mechanisms but of the very features measured or to be measured. Then any modelling assumptions of the data generation would be necessarily rather arbitrary and so too conclusions based on them.

Moreover in many cases the set of entities is rather unique and cannot be considered a sample from a larger population, such as the set of European countries or single malt whisky brands.

Sometimes the very concept of a cluster as a probabilistic distribution seems to not fit into a clustering goal. Look, for example, at a bell-shaped Gaussian distribution which is considered a good approximation for such variables as the height or weight of young male individuals of the same ethnicity so that they form a cluster corresponding to the distribution. However, when confronted with the practical issue of dividing people, for example, according to their fighting capabilities (such as in military conscription or in the sport of boxing), the set cannot be considered a homogeneous cluster anymore and must be further partitioned into more homogeneous strata. Some say that there must be a boundary between "natural" clusters and clusters to be drawn on purpose; that a bell-shape distribution corresponds to a natural cluster and a boxing weight category to an artificial one. However, it is not always easy to distinguish which situation is which. There will always be situations when a cluster of potentially weak fighters (or bad customers, or homologous proteins) must be cut out from the rest.

Machine learning perspective

Machine learning tends to view the data as a device for learning how to predict pre-specified or newly created categories. The entities are considered as coming one at a time so that the machine can learn adaptively in a supervised manner. To theorize, the flow of data must be assumed to come from a probabilistic population, an assumption which has much in common with the statistics approach. However, it is prediction rather than model fitting which is the central issue in machine learning.

Such a shift in the perspective has led to the development of strategies for predicting categories such as decision trees and support vector machines as well as resampling methods such as the bootstrap and cross-validation for dealing with limited data sets.

Data mining perspective

Data mining is not much interested in reflection on where the data have come from nor how they have been collected. It is assumed that a data set or database has been collected already and, however bad or well it reflects the properties of the phenomenon in question, the major concern is in finding patterns and regularities within the data as they are. Machine learning and statistics methods are welcome here – for their capacity to do the job.

This view, started as early as in the sixties and seventies in many countries including France, Russia and Japan in such subjects as analysis of questionnaires or of inter-industrial transfers, was becoming more and more visible, but it did not make it into prominence until the nineties. By that time, big warehouse databases became available, which led to the discovery of patterns of transactions with the so-called association search methods. The patterns proved themselves correct when superstores increased profits by accommodating to them.

Data mining is a huge activity on the intersection of databases and data analysis methods. Clustering is a recognized part of it. The data recovery approach which is maintained in this book obviously fits within data mining very well, because it is based only on the data available.

It should be added that the change of the paradigm from modeling of mechanisms of data generation to data mining has drastically changed requirements to methods and programs. According to the statistics approach, the user must know the models and methods he uses; if a method is applied wrongly, the results can be wrong too. Thus, application of statistical methods is limited within a small circle of experts. In data mining, it is the patterns not methods that matter. This shifts the focus of computer programs from statistics to the user's substantive area and makes them user-friendly.

Similarly, the validation objectives seem to diverge here: in statistics and machine learning the stress goes on the consistency of the algorithms, which is not quite so important in data mining, in which it is the consistency of patterns, not algorithms, which matters the most.

Classification/knowledge-discovery perspective

The classification perspective is rarely discussed indeed. In data mining the term "classification" is usually referred to in a very limited sense: as an activity of assigning prespecified categories (classes) to entities, in contrast to clustering which assigns entities with newly created categories (clusters).

According to its genuine meaning, classification is an actual or ideal arrangement of entities under consideration in classes to:

(1) shape and keep knowledge;

(2) capture the structure of phenomena; and

(3) relate different aspects of a phenomenon in question to each other.

These make the concept of classification a specific mechanism for knowledge discovery and maintenance. Consider, for instance, the Periodic Chart of chemical elements. Its rows correspond to numbers of electron shells in the atoms, and its columns to the numbers of electrons in the external shell thus capturing the structure of the phenomenon. These also relate to most important physical properties and chemical activities of the elements thus associating different aspects of the phenomenon. And this is a compact form of representing the knowledge; moreover, historically it is this form itself, developed rather empirically, that made possible rather fast progress to the current theories of the matter.

In spite of the fact that the notion of classification as part of scientific knowledge was introduced by the ancient Greeks (Aristotle and the like) the very term "classification" seems a missing item in the vocabulary of current scientific discourse. This may have happened because in traditional sciences, classifications are defined within well developed substantive theories according to variables which are defined as such within the theories. Thus, there has been no need in specific theories for classification.

Clustering should be considered as classification based on empirical data in a situation when clear theoretical concepts and definitions are absent and the regularities are unknown. Thus, the clustering goals should relate to the classification goals above. This brings one more aspect to clustering. Consider, for example, how one can judge whether a clustering is good or bad? According to the classification/knowledge-discovery view, this is easy and has nothing to do with statistics: just look at how well clusters fit within the existing knowledge, how well they allow updating, correcting and extending.

Somewhat simplistically, one might say that two of the points stressed in this book, that of the data recovery approach and the need to not only find, but describe clusters, fit well into the two perspectives, the former into data mining and the latter into classification as knowledge discovery.

Chapter 2

What Is Data

After reading through this chapter, the reader will know of:

1. Three types of data tables: (a) feature-to-entity, (b) similarity/dissimilarity and (c) contingency/flow tables, and ways to standardize them.
2. Quantitative, categorical and mixed data, and ways to pre-process and standardize them.
3. Characteristics of feature spread and centrality.
4. Bi-variate distributions over mixed data, correlation and association, and their characteristics.
5. Visualization of association in contingency tables with Quetelet coefficients.
6. Multidimensional concepts of distance and inner product.
7. The concept of data scatter.

Base words

Average The average value of a feature over a subset of entities. If the feature is binary and corresponds to a category, the average is the category frequency in the subset. The average over the entire entity set is referred to as a grand mean.

Contingency coefficient A summary index of statistical association between two sets of categories in a contingency table. The greater it is, the closer the association to a conceptual one.

Contingency table Given two sets of categories corresponding to rows and columns, respectively, this table presents counts of entities co-occurring at the intersection of each pair of categories from the two sets. When categories within each of the sets are mutually disjoint, the contingency table can be aggregated by summing up relevant entries.

Correlation The shape of a scatter-plot showing the extent to which two features can be considered mutually related. The (product-moment) correlation coefficient captures the extent at which one of the features can be expressed as a linear function of the other.

Data scatter The sum of squared entries of the data matrix; it is equal to the sum of feature contributions or the summary distance from entities to zero.

Data table Also referred to as *flat file* (in databases) or *vector space data* (in information retrieval), this is a two-dimensional array whose rows correspond to entities, columns to features, and entries to feature values at entities.

Distance Given two vectors of the same size, the (Euclidean squared) distance is the sum of squared differences of corresponding components, $d(x,y) = \sum_i (x_i - y_i)^2$. It is closely related to the inner product: $d(x,y) = (x - y, x - y)$.

Entity Also referred to as *observation* (in statistics) or *case* (in social sciences) or *instance* (in artificial intelligence) or *object*, this is the main item of clustering corresponding to a data table row.

Feature Also referred to as *variable* (in statistics) or *character* (in biology) or *attribute* (in logic), this is another major data item corresponding to a data table column. It is assumed that feature values can be compared to each other, at least, whether they coincide or not (categorical features), or even averaged over any subset of entities (quantitative feature case).

Inner product Given two vectors of the same size, the inner product is the sum of products of corresponding components, $(x,y) = \sum_i x_i y_i$. It is closely related to the distance: $d(x,y) = (x,x) + (y,y) - 2(x,y)$.

Quetelet index In contingency tables: A value showing the change in frequency of a row category when a column category becomes known. The greater the value, the greater the association between the column and row categories. It is a basic concept in contingency table analysis.

Range The interval in which a feature takes its values; the difference between the feature maximum and minimum over a data set.

Scatter plot A graph presenting entities as points on the plane formed by two quantitative features.

Variance The average of squared deviations of feature values from the average.

2.1 Feature characteristics

2.1.1 Feature scale types

The Masterpieces data in Table 1.10 will be used to illustrate data handling concepts in this section. For the reader's convenience, the table is reprinted here as Table 2.1.

A data table of this type represents a unity of the set of rows, always denoted as I further on, the set of columns denoted by V and the table contents X, the set of values x_{iv} in rows $i \in I$ and columns $v \in V$. The number of rows, or cardinality of I, $|I|$, will be denoted by N, and the number of columns, the cardinality of V, $|V|$, by M. Rows will always correspond to entities, columns to features. Whatever metadata of entities may be known, are all to be put as the features, except for names, that may be maintained as a list associated with I. As to the features $v \in V$, it is assumed that each has a measurement scale assigned to it, and of course a name.

All within-column entries are supposed to have been measured in the same scale and thus comparable within the scale; this is not so over rows in Y. Three different types of scales that are present in Table 2.1 and will be dealt with in the remainder are quantitative (LenSent, LenDial, and NChar), nominal (Narrative) and binary (SCon). Let us elaborate on these scale types:

Table 2.1: **Masterpieces:** Masterpieces of 19th century: the first three by Charles Dickens (1812–1870), the next three by Mark Twain (1835–1910), and the last two by Leo Tolstoy (1828–1910).

Title	LenSent	LenDial	NChar	SCon	Narrative
Oliver Twist	19.0	43.7	2	No	Objective
Dombey and Son	29.4	36.0	3	No	Objective
Great Expectations	23.9	38.0	3	No	Personal
Tom Sawyer	18.4	27.9	2	Yes	Objective
Huckleberry Finn	25.7	22.3	3	Yes	Personal
Yankee at King Arthur	12.1	16.9	2	Yes	Personal
War and Peace	23.9	30.2	4	Yes	Direct
Anna Karenina	27.2	58.0	5	Yes	Direct

1. **Quantitative**: A feature is quantitative if the operation of taking its average is meaningful.

   It is quite meaningful to compare the average values of feature LenS or LenD for different authors in Table 2.1. Somewhat less convincing is the case of NumC which must be an integer; some authors even consider such "counting" features a different scale type. Still, we can safely say that on average Tolstoy's novels have larger numbers of principal characters than those by Dickens or Twain. This is why counting features are also considered quantitative in this text.

2. **Nominal**: A categorical feature is said to be nominal if its categories are (i) disjoint, that is, no entity can fall in more than one of them, and (ii) not ordered, that is, they only can be compared with respect to whether they coincide or not. Narrative, in Table 2.1, is such a feature.

   Categorical features maintaining (ii) but not (i) are referred to as multi-choice variables. For instance, Masterpieces data might include a feature that presents a list of social themes raised in a novel, which may contain more than one element. That would produce a one-to-many mapping of the entities to the categories, that is, social themes. There is no problem in treating this type of data within the framework described here. For instance, the Digit data table may be treated as that representing the only, multi-choice, variable "Segment" which has the set of seven segments as its categories.

   Categorical features that maintain (i) but have their categories ordered are called rank variables. Variable Bribe level in the Bribery data of Tables 1.11 and 1.12 is rank: its three categories are obviously ordered according to the bribe size. Traditionally, it is assumed for the rank variables that only the order of categories matters and intervals between them are irrelevant. That is, rank categories may accept any quantitative coding which is compatible with their ordering. This makes rank features difficult to deal with in the context of mixed data tables. We maintain a different view, going back to C. Spearman: the ranks are treated as numerical values and the rank variables are considered thus quantitative and processed accordingly. In particular, seven of the eleven variables in Bribery data (II. Client, IV. Occurrence, V. Initiator, VI. Bribe, VII. Type, VIII. Network, and XI. Punishment) will be considered ranked with ranks assigned in Table 1.11 and treated as quantitative values.

   There are two approaches to the issue of involving qualitative features into analysis. According to one, more traditional, approach, categorical variables are considered non-treatable quantitatively. The only quantitative operation admitted for categories is counting the number or frequency of its occurrences at various subsets. To conduct cluster analysis, categorical data, according to this view, can only be utilized for deriving an entity-to-entity (dis)similarity measure. Then this measure can be used for finding clusters.

   A different approach is maintained and further developed here: a category defines a quantitative zero-one variable on entities, with one corresponding to its presence and zero absence, which is treated then as such. We will see later that this view, in fact, does not contradict the former one but rather fits into it with geometrically and statistically sound specifications.

3. **Binary**: A qualitative feature is said to be binary if it has two categories which can be thought of as Yes or No answer to a question such as feature SCon in Table 2.1. A two-category feature can be considered either a nominal or binary one, depending on the context. For instance, feature "Gender" of a human should be considered a nominal feature, whereas the question "Are you female?" a binary feature, because the latter assumes that it is the "female," not "male," category which is of interest. Operationally, the difference between these two types will amount to how many binary features should be introduced to represent the feature under consideration in full. Feature "Gender" cannot be represented by one column with Yes or No categories: two are needed, one for "Female" and one for "Male."

2.1.2 Quantitative case

As mentioned, we consider that the meaningfulness of taking the average is a defining property of a quantitative variable. Given a feature $v \in V$ whose values y_{iv}, $i \in I$, constitute a column in the data table, its *average* over entity subset $S \subseteq I$ is defined by the formula

$$c_v(S) = (1/N_S) \sum_{i \in S} y_{iv} \tag{2.1}$$

where N_S is the number of entities in S.

The average $c_v = c_v(I)$ of $v \in V$ over the entire set I is sometimes referred to as *grand mean*. After grand mean c_v of $v \in V$ has been subtracted from all elements of the column-feature $v \in V$, the grand mean of v becomes zero. Such a variable is referred to as *centered*.

It should be mentioned that usually the quantitative scale is defined somewhat differently, not in terms of the average but the so-called admissible transformations $y = \phi(x)$. The scale type is claimed to depend on the set of transformations ϕ which are considered admissible, that is, do not change the scale contents. For the quantitative feature scales, those that are admissible are transformations such as $y = ax + b$ converting all x values into y values by changing the scale factor a times and shifting the scale origin at b. Transformations $y = \phi(x)$ of this type, with $\phi(x) = ax+b$ for some real a and b, are referred to as affine transformations. For instance, the temperature Celsius scale x is transformed into the temperature Fahrenheit scale with $\phi(x) = 1.8x+32$. Standardizations of data with affine transformations are at the heart of our approach to clustering.

Our definition is compatible with the one given above. Indeed, if a feature x admits affine transformations, it is meaningful to compare its average values over various entity sets. Let x_J and x_K be the averages of sets $\{x_j : j \in J\}$ and $\{x_k : k \in K\}$ respectively, and, say, $x_J \leq x_K$. Does the same automatically hold for the averages of $y = ax + b$ over J and K? To answer this question, we consider values $y_j = ax_j + b$, $j \in J$, and $y_k = ax_k + b$, $k \in K$ and calculate their averages, y_J and y_K. It is easy to prove that $y_K = ax_K + b$ and $y_J = ax_J + b$ so that any relation between x_J and x_K remains the same for y_J and y_K, up to the obvious reversal when a is negative (which means that rescaling involves change of the direction of the scale).

Other indices of "centrality" have been considered too; the most popular of them are:

i Midrange, point in the middle of the range, that is, equi-distant from the minimum and maximum values of the feature.

ii Median, the middle item in the series of elements of column v sorted in ascending (or descending) order.

iii Mode, "the most likely" value, which is operationally defined by partitioning the feature range in a number of bins (intervals of the same size) and determining at which of the bins the number of observations is maximum: the center of this bin is the mode, up to the error related to the bin size.

Each of these has its advantages and drawbacks as a centrality measure. The median is the most stable with regard to change in the sample and, especially, to the presence of outliers. Outliers can drastically change the average, and they do not affect the median at all. However, the calculation of the median requires sorting the entity set, which sometimes may be costly. Midrange is insensitive to the shape of the distribution and is highly sensitive to outliers. The mode is of interest when distribution of the feature is far from uniform.

These may give a hint with respect to what measure should be used in a specific situation. For example, if the data to be analyzed have no specific properties at all, the average should be utilized. When outliers or data errors are expected, the median would be a better bet.

The average, median and midrange all fit within the following approximation model which is at the heart of the data recovery approach. Given a number of reals, x_1, x_2,..., x_N, find a unique real a that can be used as their aggregate substitute so that for each i, a approximates x_i up to a residual ϵ_i: $x_i = a + \epsilon_i$, $i \in I$. The smaller the residuals the better the aggregate. To minimize the residuals $\epsilon_i = x_i - a$, they should be combined into a scalar criterion such

Table 2.2: Summary characteristics of the Market town data.

	P	PS	Do	Ho	Ba	Su	Pe	DIY	SP	PO	CAB	FM
Mean	7351.4	3.0	1.4	0.4	4.3	1.9	2.0	0.2	0.5	2.6	0.6	0.2
Std	6193.2	2.7	1.3	0.6	4.4	1.7	1.6	0.6	0.6	2.1	0.6	0.4
Range	21761.0	12.0	4.0	2.0	19.0	7.0	7.0	3.0	2.0	8.0	2.0	1.0

as $L_1 = \sum_i |x_i - a|$, $L_\infty = \max_i |x_i - a|$, or $L_2 = \sum_i |x_i - a|^2$. It appears, L_1 is minimized by the median, L_∞ by midrange and L_2 by the average. The average fits best because it solves the least squares approximation problem, and the least-square criterion is the basis of all further developments in Chapter 5.

A number of characteristics have been defined to measure the features' dispersion or spread. Probably the simplest of them is the variable's *range*, the difference between its maximal and minimal values, that has been mentioned above already. This measure should be used cautiously as it may be overly sensitive to changes in the entity set. For instance, removal of Truro from the set of entities in the Market town data immediately reduces the range of variable Banks to 14 from 19. Further removal of St Blazey/Par further reduces the range to 13. Moreover, the range of variable DIY shrinks to 1 from 3, with these two towns removed. Obviously, no such drastic changes emerge when all thirteen hundred of the English Market towns are present.

A somewhat more elaborate characteristic of dispersion is the so-called (empirical) variance of $v \in V$ which is defined as

$$s_v^2 = \sum_{i \in I} (y_{iv} - c_v)^2 / N \tag{2.2}$$

where c_v is the grand mean. That is, s_v^2 is the average squared deviation L_2 of y_{iv} from c_v.

The standard deviation of $v \in V$ is defined as just $s_v = \sqrt{s_v^2}$ which has also a statistical meaning as the square-average deviation of the variable's values from its grand mean. The standard deviation is zero, $s = 0$, if and only if the variable is constant, that is, all the entries are equal to each other.

In some packages, especially statistical ones, denominator $N - 1$ is used instead of N in definition (2.2) because of probabilistic consistency considerations (see any text on mathematical statistics). This shouldn't much affect results because N is assumed constant here and, moreover, $1/N$ and $1/(N-1)$ do not much differ when N is large.

For the Market town data, with $N = 45$ and $n = 12$, the summary characteristics are in Table 2.2.

The standard deviations in Table 2.2 are at least as twice as small as the ranges, which is true for all data tables (see Statement 2.2.).

The values of the variance s_v^2 and standard deviation s_v obviously depend on the variable's spread measured by its range. Multiplying the column $v \in V$

by $\alpha > 0$ obviously multiplies its range and standard deviation by α, and the variance by α^2.

The quadratic index of spread, s_v^2, depends not only on the scale but also on the character of the feature's distribution within its range. Can we see how?

Let as consider all quantitative variables defined on N entities $i \in I$ and ranged between 0 and 1 inclusive, and analyze at what distributions the variance attains its maximum and minimum values.

It is not difficult to see that any feature v that minimizes the variance s_v^2 is equal to 0 at one of the entities, 1 at another entity, and $y_{iv} = c_v = 1/2$ at all other entities $i \in I$. The minimum variance s_v^2 is $\frac{1}{2N}$ then.

Among the distributions under consideration, the maximum value of s_v^2 is reached at a feature v which is binary, that is, has only boundary points, 0 and 1, as its values. Indeed, if v has any other value at an entity i, then the variance will only increase if we redefine v in such a way that it becomes 0 or 1 at i depending on whether y_{iv} is smaller or greater than v's average c_v. For a binary v, let us specify proportion p of values y_{iv} at which the feature is larger than its grand mean, $y_{iv} > c_v$. Then, obviously, the average $c_v = 0 * (1-p) + 1 * p = p$ and, thus, $s_v^2 = (0-p)^2 * (1-p) + (1-p)^2 * p = p(1-p)$.

The choice of the left and right bounds of the range, 0 and 1, does have an effect on the values attained by the extremal variable but not on the conclusion of its binariness. That means that the following is proven.

Statement 2.1. *With the range and proportion p of values smaller than the average prespecified, the distribution at which the variance reaches its maximum is the distribution of a binary feature having p values at the left bound and $1-p$ values at the right bound of the range.*

Among the binary features, the maximum variance is reached at $p = 1/2$, the maximum uncertainty. This implies one more property.

Statement 2.2. *For any feature, its standard deviation is at least as twice as small as its range.*

Proof: Indeed, with the range being unity between 0 and 1, the maximum variance is $p(1-p) = 1/4$ at $p = 1/2$ leading to the maximum standard deviation of just half of the unity range, q.e.d.

From the intuitive point of view, the range being the same, the greater the variance the better the variable suits the task of clustering.

2.1.3 Categorical case

Let us first consider binary features and then nominal ones.

To quantitatively recode a binary feature, its Yes category is converted into 1 and No into 0. The grand mean of the obtained zero/one variable will be

p_v, the proportion of entities falling in the category Yes. Its variance will be $s_v^2 = p(1-p)$.

In statistics, two types of probabilistic mechanisms for generating zero/one binary variables are considered, Bernoulli/binomial and Poisson. Each relies on having the proportion of ones, p, fixed. However, the binomial distribution assumes that every single entry has the same probability p of being unity, whereas Poisson distribution does not care about individual entries: just that the proportion p of entries randomly thrown into a column must be unity. This subtle difference makes the variance of the Poisson distribution greater: the variance of the binomial distribution is equal to $s_v^2 = p(1-p)$ and the variance of the Poisson distribution is equal to $\pi_v = p$. Thus, the variance of a one-zero feature considered as a quantitative feature corresponds to the statistical model of binomial distribution.

Turning to the case of nominal variables, let us denote the set of categories of a nominal variable l by V_l. Any category $v \in V_l$ is conventionally characterized by its frequency, the number of entities, N_v, falling in it. The sum of frequencies is equal to the total number of entities in I, $\sum_{v \in V_l} N_v = N$. The *relative frequencies*, $p_v = N_v/N$, sum up to unity. The vector $p = (p_v)$, $v \in V_l$ is referred to as the *distribution* of l (over I). A category with the largest frequency is referred to as the distribution's mode. The dispersion of a nominal variable l is frequently measured by the so-called *Gini coefficient*, or qualitative variance:

$$G = \sum_{v \in V_l} p_v(1 - p_v) = 1 - \sum_{v \in V_l} p_v^2 \tag{2.3}$$

This is zero if all entities fall in one of the categories only. G is maximum when the distribution is *uniform*, that is, when all category frequencies are the same, $p_v = 1/|V_l|$ for all $v \in V_l$.

A similar measure referred to as *entropy* and defined as

$$H = -\sum_{v \in V_l} p_v \log p_v \tag{2.4}$$

with the logarithm's base 2 is also quite popular. This measure is related to so-called information theory [12]. Entropy reaches its minimum and maximum values at the same distributions as the Gini coefficient. Moreover, the Gini coefficient can be thought of as a linearized version of entropy since $1 - p_v$ linearly approximates $\log p_v$ at p_v close to 1. In fact, both can be considered averaged information measures, just that one uses $-\log p_v$ and the other $1 - p_v$ to express the information contents.

There exists a general formula to express the diversity of a nominal variable as $S_q = (1 - \sum_{v \in V_l} p_v^q)/(q-1)$, $q > 0$ [132]. The entropy and Gini index are special cases of S_q since $S_2 = G$ and $S_1 = H$ assuming S_1 to be the limit of S_q when q tends to 1.

A nominal variable l can be converted into a quantitative format by assigning a zero/one feature to each of its categories $v \in V_l$ coded by 1 or 0 depending on whether an entity falls into the category or not. These binary features are referred to sometimes as dummy variables.

Unlike a binary feature, a two-category nominal feature such as "Gender" is converted into two columns, each corresponding to one of the categories, "Male" and "Female" of "Gender." This way of quantization is quite convenient within the data recovery approach as will be seen further in section 4.3 and others. However, it is also compatible with the traditional view of quantitative measurement scales as expressed in terms of admissible transformations. Indeed, for a nominal scale x, any one-to-one mapping $y = \phi(x)$ is considered admissible. When there are only two categories, x_1 and x_2, they can be recoded into any y_1 and y_2 with an appropriate rescaling factor a and shift b so that the transformation of x to y can be considered an affine one, $y = ax + b$. It is not difficult to prove that $a = (y_1 - y_2)/(x_1 - x_2)$ and $b = (x_1y_2 - x_2y_1)/(x_1 - x_2)$ will do the recoding. In other words, nominal features with two categories can be considered quantitative. The binary features, in this context, are those with category Yes coded by 1 and No by 0 for which transformation $y = ax + b$ is meaningful only when $a > 0$.

The vector of averages of the dummy category features, p_v, $v \in V_l$, is nothing but the distribution of l. Moreover, the Gini coefficient appears to be but the summary Bernoullian variance of the dummy category features, $G = \sum_{v \in V_l} p_v(1 - p_v)$. In the case when l has only two categories, this becomes just the variance of any of them doubled. Thus, the transformation of a nominal variable into a bunch of zero-one dummies conveniently converts it into a quantitative format which is compatible with the traditional treatment of nominal features.

2.2 Bivariate analysis

Statistical science in the pre-computer era developed a number of tools for the analysis of interrelations between variables, which will be useful in the sequel. In the remainder of this section, a review is given of the three cases emerging from the pair-wise considerations, with emphasis on the measurement scales: (a) quantitative-to-quantitative, (b) categorical-to-quantitative, and (c) categorical-to-categorical variables. The discussion of the latter case follows that in [93].

2.2.1 Two quantitative variables

Mutual interrelations between two quantitative features can be caught with a scatter plot such as in Figure 1.10, page 24. Two indices for measuring

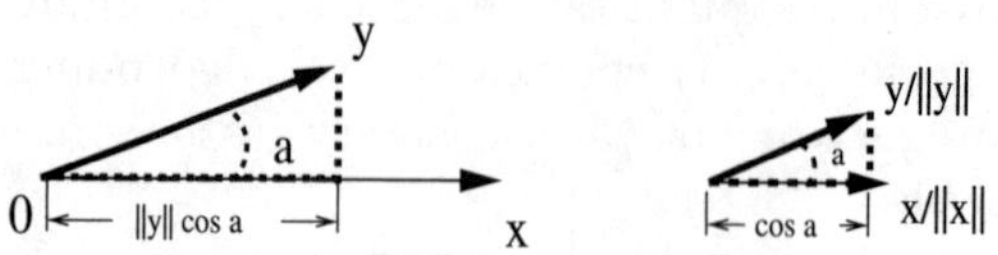

Figure 2.1: Geometrical meaning of the inner product and correlation coefficient.

association between quantitative variables have attracted considerable attention in statistics and data mining: those of covariance and correlation.

The covariance coefficient between the variables x and y considered as columns in a data table, $x = (x_i)$ and $y = (y_i)$, $i \in I$, can be defined as

$$cov(x, y) = (1/N) \sum_{i \in I} (x_i - \bar{x})(y_i - \bar{y}) \quad (2.5)$$

where $\bar{x}$ and $\bar{y}$ are the average values of x and y, respectively.

Obviously, $cov(x, x) = s^2(x)$, the variance of x defined in section 2.1.3.

The covariance coefficient changes proportionally when the variable scales are changed. A scale-invariant version of the coefficient is the correlation coefficient (sometimes referred to as the Pearson product-moment correlation coefficient) which is the covariance coefficient normalized by the standard deviations:

$$r(x, y) = cov(x, y)/(s(x)s(y)) \quad (2.6)$$

A somewhat simpler formula for the correlation coefficient can be obtained if the data are first standardized by subtracting their average and dividing the results by the standard deviation: $r(x, y) = cov(x', y') = (x', y')/N$ where $x'_i = (x_i - \bar{x})/s(x)$, $y'_i = (y_i - \bar{y})/s(y)$, $i \in I$. Thus, the correlation coefficient is but the mean of the component-to-component, that is, inner, product of feature vectors when both of the scales are standardized as above.

The coefficient of correlation can be substantiated in different theoretic frameworks. These require some preliminary knowledge of mathematics and can be omitted at first reading, which is reflected in using a smaller font for explaining them.

1. **Cosine.** A geometric approach, relying on concepts introduced later in section 2.3.2, offers the view that the covariance coefficient as the inner product of feature column-vectors is related to the angle between the vectors so that $(x, y) = ||x|| ||y|| \cos(x, y)$. This can be illustrated with Fig. 2.1; norms $||x||$ and $||y||$ are Euclidean lengths of intervals from 0 to x and y, respectively. The correlation coefficient is the inner product of the corresponding normalized variables, that is, the cosine of the angle between the vectors.

2. **Linear slope.** The data recovery approach suggests that one of the features is modeled as a linear function of the other, say, y as $ax + b$ where a and b are chosen to minimize the norm of the difference, $||y - ax - b||$. It appears, the optimal slope a is proportional to $r(x, y)$ and, moreover, the square $r(x, y)^2$ expresses that part of the variance of y that is taken into account by $ax + b$ (see details in section 5.1.2).
3. **Parameter in Gaussian distribution.** The correlation coefficient has a very clear meaning in the framework of probabilistic bivariate distributions. Consider, for the sake of simplicity, features x and y normalized so that the variance of each is unity. Denote the matrix formed by the two features by $z = (x, y)$ and assume a unimodal distribution over z, controlled by the so-called Gaussian, or normal, density function (see section 6.1.5) which is proportional to the exponent of $-z^T\Sigma^{-1}z/2$ where Σ is a 2×2 matrix equal to $\Sigma = \begin{pmatrix} 1 & r \\ r & 1 \end{pmatrix}$. The parameter r determines the distance between the foci of the ellipse $z^T\Sigma^{-1}z = 1$: the greater r the greater the distance. At $r = 0$ the distance is zero so that the ellipsis is a circle and at r tending to 1 or -1 the distance tends to the infinity so that the ellipse degenerates into a straight line. It appears $r(x, y)$ is a sample based estimate of this parameter.

These three frameworks capture different pieces of the "elephant." That of cosine is the most universal framework: one may always take that measure to see to what extent two features go in concert, that is, to what extent their highs and lows co-occur. As any cosine, the correlation coefficient is between –1 and 1, the boundary values corresponding to the coincidence of the normalized variables or to a linear relation between the original features. The correlation coefficient being zero corresponds to the right angle between the vectors: the features are not correlated! Does that mean they must be independent in this case? Not necessarily. The linear slope approach allows one to see how this may happen: just the slope of the line best fitting the scatter-plot must be horizontal. According to this approach, the square of the correlation coefficient shows to what extent the relation between variables, as observed, is owed to linearity. The Gaussian distribution view is the most demanding: it requires a properly defined distribution function, a unimodal one, if not normal.

Three examples of scatter-plots on Figure 2.2 illustrate some potential cases of correlation: (a) strong positive, (b) strong negative, and (c) zero correlation. Yet be reminded: in contrast to what is claimed in popular web sites, the the correlation coefficient cannot gather up all the cases in which variables are related; it does capture only those of linear relation and those close enough to that.

2.2.2 Nominal and quantitative variables

How should one measure association between a nominal feature and a quantitative feature? By looking at whether specifying a category can lead to a better

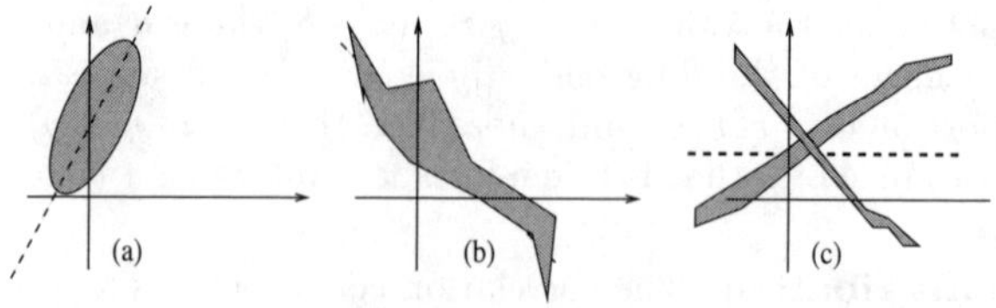

Figure 2.2: Three cases of a scatter-plot: (a) positive correlation, (b) negative correlation, (c) no correlation. The shaded area is supposed to be randomly covered by entity points.

prediction of the quantitative feature or not.

Let us denote the partition of the entity set I corresponding to categories of the nominal variable by $S = \{S_1, ..., S_m\}$; subset S_k consists of N_k entities falling in k-th category of the variable. The quantitative variable will be denoted by y with its values y_i for $i \in I$. The box-plot such as in Figure 1.7 on page 23 is a visual representation of the relationship between the nominal variable represented by the grouping and the quantitative variable represented by the boxes and whiskers.

Let us introduce the framework for prediction of y values. Let the predicted y value for any entity be the grand mean $\bar{y}$ if no other information is supplied, or $\bar{y}_k = \sum_{i \in S_k} y_i/N_k$, the within-class average, if the entity is known to belong to S_k. The average error of these predictions can be estimated in terms of the variances. To do this, one should relate within-class variances of y to its total variance s^2: the greater the change, the lesser the error and the closer the relation between S and y.

An index, referred to as the correlation ratio, measures the proportion of total feature variance that falls within classes. Let us denote the within class variance of variable y by

$$s_k^2 = \sum_{i \in S_k} (y_i - \bar{y}_k)^2/N_k \tag{2.7}$$

where N_k is the number of entities in S_k and $\bar{y}_k$ the feature's within cluster average. Let us denote the proportion of S_k in I by p_k so that $p_k = N_k/N$. Then the average variance within partition $S = \{S_1, ..., S_m\}$ will be $\sum_{k=1}^{K} p_k s_k^2$. This can be proven to never be greater than the total variance s^2.

However, the average within partition variance can be as small as zero – when values y_i all coincide with $\bar{y}_k$ within each category S_k, that is, when y is piece-wise constant across S. In other words, all cluster boxes of a box-plot of y over classes of S degenerate into straight lines in this case. In such a situation partition S is said to perfectly match y. The smaller the difference between

Table 2.3: Cross-classification of 8 masterpieces according to the author and Narrative in the format Count/Proportion.

	Narrative			
Author	Objective	Personal	Direct	Total
Dickens	2/0.250	1/0.125	0/0	3/0.375
Twain	1/0.125	2/0.250	0/0	3/0.375
Tolstoy	0/0	0/0	2/0.250	2/0.250
Total	3/0.375	3/0.375	2/0.375	8/1.000

the average within-class variance and s^2, the worse the match between S and y. The relative value of the difference,

$$\eta^2 = \frac{s^2 - \sum_{k=1}^{K} p_k s_k^2}{s^2} \tag{2.8}$$

is referred to as the correlation ratio.

The correlation ratio is between 0 and 1, the latter corresponding to the perfect match case. The greater the within-category variances, the smaller the correlation ratio. The minimum value, zero, is reached when all within class variances coincide with the total variance.

2.2.3 Two nominal variables cross-classified

Interrelation between two nominal variables is represented with the so-called contingency table. A contingency, or cross-classification, data table corresponds to two sets of disjoint categories, such as authorship and narrative style in the Masterpieces data, which respectively form rows and columns of Table 2.3.

Entries of the contingency table are co-occurrences of row and column categories, that is, counts of numbers of entities that fall simultaneously in the corresponding row and column categories such as in Table 2.3.

In a general case, with the row categories denoted by $t \in T$ and column categories by $u \in U$, the co-occurrence counts are denoted by N_{tu}. The frequencies of row and column categories usually are called marginals (since they are presented on margins of contingency tables as in Table 2.3) and denoted by N_{t+} and N_{+u} since, when the categories within each of the two sets do not overlap, they are sums of co-occurrence entries, N_{tu}, in rows, t, and columns, u, respectively. The proportions, $p_{tu} = N_{tu}/N$, $p_{t+} = N_{t+}/N$, and $p_{+u} = N_{+u}/N$ are also frequently used as contingency table entries. The general contingency table is presented in Table 2.4.

Contingency tables can be considered for quantitative features too, if they are preliminarily categorized as demonstrated in the following example.

Table 2.4: A contingency table or cross classification of two sets of categories, $t \in T$ and $u \in U$ on the entity set I.

Category	1	2	...	$\|U\|$	Total
1	N_{11}	N_{12}	...	$N_{1\|U\|}$	N_{1+}
2	N_{21}	N_{22}	...	$N_{2\|U\|}$	N_{2+}
...	...	...	....	...	...
$\|T\|$	$N_{\|T\|1}$	$N_{\|T\|2}$	...	$N_{\|T\|\|U\|}$	$N_{\|T\|+}$
Total	N_{+1}	N_{+2}	...	$N_{+\|U\|}$	N

Table 2.5: Cross classification of the Bank related partition with FM feature at Market towns.

	Number of banks				
FMarket	10+	4+	2+	1-	Total
Yes	2	5	1	1	9
No	4	7	13	12	36
Total	6	12	14	13	45

Table 2.6: Frequencies, per cent, in the bivariate distribution of the Bank related partition and FM at Market towns.

	Number of banks				
FMarket	10+	4+	2+	1-	Total
Yes	4.44	11.11	2.22	2.22	20.00
No	8.89	15.56	28.89	26.67	80.00
Total	13.33	26.67	31.11	28.89	100.00

Example 2.1. A cross classification of Market towns

Let us partition the Market town set in four classes according to the number of Banks and Building Societies (feature Ba): class T_1 to include towns with Ba equal to 10 or more; T_2 with Ba equal to 4 or more, but less than 10; T_3 with Ba equal to 2 or 3; and T_4 to consist of towns with one or no bank at all. Let us cross classify partition $T = \{T_v\}$ with feature FM, presence or absence of a Farmers' market in the town. That means that we draw a table whose columns correspond to classes T_v, rows to presence or absence of Farmers' markets, and entries to their overlaps (see Table 2.5).

The matrix of frequencies, or proportions, $p_{tu} = N_{tu}/N$ for Table 2.5 can be found by dividing all its entries by $N = 45$ (see Table 2.6). □

A contingency table gives a picture of interrelation between two categorical features, or partitions corresponding to them, which is not quite clear. Let us make the picture sharper by removing thirteen towns from the sample, those

Table 2.7: Cross classification of the Bank related partition with FM feature at a "cleaned" subsample of Market towns.

	Number of banks				
FMarket	10+	4+	2+	1-	Total
Yes	2	5	0	0	7
No	0	0	13	12	25
Total	2	5	13	12	32

falling in the less populated cells of Table 2.5 (see Table 2.7). Table 2.7 shows a very clear association between two features on the "cleaned" subsample: the Farmers' markets are present only in towns in which the number of banks is 4 or greater. A somewhat subtler relation: the medium numbers of banks are more closly associated with the presence of a Farmers' market than the higher ones.

This clear picture is somewhat blurred in the original sample in Table 2.5 and, moreover, maybe does not hold at all.

Thus, the issue of relating two features to each other can be addressed by looking at mismatches. For instance, Table 2.3 shows that Narrative style is quite close to authorship, though they do not completely match: there are two mismatching entities, one by Dickens and the other by Twain. Similarly, there are 13 mismatches in Table 2.5 removed in Table 2.7. The sheer numbers of mismatching entities measure the differences between category sets rather well when the distribution of entities within each category is rather uniform as it is in Table 2.3. When the proportions of entities in different categories drastically differ, as in Table 2.5, to measure association between category sets more properly, the numbers of mismatching entities should be weighted according to the frequencies of corresponding categories. Can we discover the relation in Table 2.5 without removing entities?

To measure association between categories according to a contingency table, a founding father of the science of statistics, A. Quetelet, proposed utilizing the relative or absolute change of the conditional probability of a category. The conditional probability $p(u/t) = N_{tu}/N_{t+} = p_{tu}/p_{t+}$ measures the proportion of category u in category t. Quetelet coefficients measure the difference between $p(u/t)$ and the average rate p_{+u} of $u \in U$. The Quetelet absolute probability change is defined as

$$g_{tu} = p(u/t) - p_{+u} = (p_{tu} - p_{t+}p_{+u})/p_{t+}, \tag{2.9}$$

and the Quetelet relative change

$$q_{tu} = g_{tu}/p_{+u} = p_{tu}/(p_{t+}p_{+u}) - 1 = (p_{tu} - p_{t+}p_{+u})/(p_{t+}p_{+u}) \tag{2.10}$$

If, for instance, t is an illness risk factor such as "exposure to certain allergens" and u is an allergic reaction such as asthma, and $p_{tu} = 0.001, pt+ = 0.01, p_{+u} = 0.02$, that means that ten per cent of those people who have been exposed to the allergens, $p(u/t) = p_{tu}/p_{t+} = 0.001/0.1 = 0.1$, contract the disease while only two per cent on average have the disease. Thus, the exposure to the allergens multiplies risk of the disease fivefold or increases the probability of contracting it by 400 %. This is exactly the value of $q_{tu} = 0.001/0.0002 - 1 = 4$. The value of g_{tu} expresses the absolute difference between $p(u/t) = 0.1$ and $p_{+u} = 0.02$; it is not that dramatic, just 0.08.

The summary Quetelet coefficients (weighted by the co-occurrence values) can be considered as summary measures of association between two category sets especially when distributions are far from uniform:

$$G^2 = \sum_{t \in T} \sum_{u \in U} p_{tu} g_{tu} = \sum_{t \in T} \sum_{u \in U} \frac{p_{tu}^2}{p_{t+}} - \sum_{u \in U} p_{+w}^2 \tag{2.11}$$

and

$$Q^2 = \sum_{t \in T} \sum_{u \in U} p_{tu} q_{tu} = \sum_{t \in T} \sum_{u \in U} \frac{p_{tu}^2}{p_{t+} p_{+u}} - 1 \tag{2.12}$$

The right-hand 1 in (2.12) comes as $\sum_{t \in T} \sum_{u \in U} p_{tu}$ when the categories t are mutually exclusive and cover the entire set I as well as categories u. In this case Q^2 is equal to the well known Pearson chi-squared coefficient X^2 defined by a different formula:

$$X^2 = \sum_{t \in T} \sum_{u \in U} \frac{(p_{tu} - p_{t+} p_{+u})^2}{p_{t+} p_{+u}} \tag{2.13}$$

The fact that $Q^2 = X^2$ can be proven with simple algebraic manipulations. Indeed, take the numerator in X^2: $(p_{tu} - p_{t+} p_{+u})^2 = p_{tu}^2 - 2p_{tu} p_{t+} p_{+u} + p_{t+}^2 p_{+u}^2$. Divided by the denominator $p_{t+} p_{+u}$, this becomes $p_{tu}^2 / p_{t+} p_{+u} - 2p_{tu} + p_{t+} p_{+u}$. Summing up the first item over all u and t leads to $Q^2 + 1$. The second item sums up to -2 and the third item to 1, which proves the statement.

This coefficient is by far the most popular association coefficient. There is a probability-based theory describing what values of NX^2 can be explained by fluctuations of the random sampling from the population.

The difference between equivalent expressions (2.12) and (2.13) for the relative Quetelet coefficient Q^2 reflects deep epistemological differences. In fact, Pearson chi-squared coefficient has been introduced in the format of NX^2 with X^2 in (2.13) to measure the deviation of the bivariate distribution in an observed contingency table from the model of statistical independence. Two partitions (categorical variables) are referred to as statistically independent if any entry in their relative contingency table is equal to the product of corresponding marginal proportions; that is, in our notation,

$$p_{tu} = p_{t+}p_{+u} \tag{2.14}$$

for all $t \in T$ and $u \in U$.

Expression (2.13) for X^2 shows that it is a quadratic measure of deviation of the contingency table entries from the model of statistical independence. This shows that (2.13) is good at testing the hypothesis of statistical independence when I is an independent random sample: the statistical distribution of NX^2 has been proven to converge, when N tends to infinity, to the chi-squared distribution with $(|T|-1)(|U|-1)$ degrees of freedom.

Statistics texts and manuals claim that, without relating the observed contingency counts to the model of statistical independence, there is no point in considering X^2. This claim sets a very restrictive condition for using X^2 as an association measure. In particular, it is quite cumbersome to substantiate presence of zero entries (such as in Tables 2.3 and 2.7) in a contingency table under this condition. However, expression (2.12) for Q^2 sets a very different framework that has nothing to do with the statistical independence. In this framework, X^2 is Q^2, the average relative change of the probability of a category u when category t becomes known. There is no restriction on using $X^2 = Q^2$ in this framework.

It is not difficult to prove that the summary coefficient Q^2 reaches its maximum value

$$\max X^2 = \min(|U|, |T|) - 1 \tag{2.15}$$

in tables with the structure of Table 2.7, at which only one element is not zero in every column (or row, if the number of rows is greater than the number of columns)[93]. Such a structure suggests a conceptual relation between categories of the two features, which means that the coefficient is good in measuring association indeed. For instance, according to Table 2.7, Farmers' markets are present if and only if the number of banks or building societies is 4 or greater, and $Q^2 = 1$ in this table.

The minimum value of Q^2 is reached in the case of statistical independence between the features, which obviously follows from the "all squared" form of the coefficient in (2.13).

Formula (2.12) suggests a way for visualization of dependencies in a "blurred" contingency table by putting the constituent items $p_{tu}q_{tu}$ as (t,u) entries of a show-case table. The proportional but greater values $Np_{tu}q_{tu} = N_{tu}q_{tu}$ can be used as well, since they sum up to NX^2 used in the probabilistic framework.

Table 2.8: Relative Quetelet coefficients, per cent, for Table 2.5.

	Number of banks			
FMarket	10+	4+	2+	1-
Yes	**66.67**	**108.33**	-64.29	-61.54
No	-16.67	-27.08	**16.07**	**15.38**

Table 2.9: Items summed up in the chi-square contingency coefficient (times N) in the Quetelet format (2.12) for Table 2.5.

	Number of banks				
FMarket	10+	4+	2+	1-	Total
Yes	**1.33**	**5.41**	-0.64	-0.62	5.48
No	-0.67	-1.90	**2.09**	**1.85**	1.37
Total	0.67	3.51	1.45	1.23	6.86

Example 2.2. Highlighting positive contributions to the total association

The table of the relative Quetelet coefficients q_{tu} for Table 2.5 is presented in Table 2.8 and that of items $N_{tu}q_{tu}$ in Table 2.9.

It is easy to see that the highlighted positive entries in both of the tables express the same pattern as in Table 2.7 but without removing entities from the table.

Table 2.9 demonstrates one more property of the items $p_{tu}q_{tu}$ summed up in the chi-square coefficient: their within-row or within-column sums are always positive. □

Highlighting the positive entries $p_{tu}q_{tu} > 0$ (or $q_{tu} > 0$) can be used for visualization of the pattern of association between any categorical features [94].

A similar to (2.13), though asymmetric, expression can be derived for G^2:

$$G^2 = \sum_{t \in T} \sum_{u \in U} \frac{(p_{tu} - p_{t+}p_{+u})^2}{p_{t+}} \tag{2.16}$$

Though it also can be considered a measure of deviation of the contingency table from the model of statistical independence, G^2 has been always considered in the literature as a measure of association. A corresponding definition involves the Gini coefficient defined in section 2.1.3, $G(U) = 1 - \sum_{u \in U} p_{+u}^2$. Within a category t, the variation is equal to $G(U/t) = 1 - \sum_{u \in U}(p_{tu}/p_{t+})^2$, which makes, on average, the qualitative variation that cannot be explained by T: $G(U/T) = \sum_{t \in T} p_{t+}G(U/t) = 1 - \sum_{t \in T}\sum_{u \in U} p_{tu}^2/p_{t+}$.

The difference $G(U) - G(U/T)$ represents that part of $G(U)$ that is explained by T, and this is exactly G^2 in (2.11).

2.2.4 Relation between correlation and contingency

Let us elaborate on the interrelation between the correlation and contingency. K. Pearson tackled the issue by proving that, given two quantitative features whose ranges have been divided into a number of equal intervals, under some standard mathematical assumptions, the value of $\sqrt{X^2/(1+X^2)}$ converges to the correlation coefficient when the number of intervals tends to infinity [63].

To define a framework for experimentally exploring the issue in the context of a mixed scale pair of features, let us consider a quantitative feature A and a nominal variable A_t obtained by partitioning the range of A into t qualitative categories, with respect to a pre-specified partition $S = \{S_k\}$. The relation between S and A can be captured by comparing the correlation ratio $\eta^2(S, A)$ with corresponding values of contingency coefficients $G^2(S, A_t)$ and $Q^2(S, A_t)$. The choice of the coefficients is not random. As proven in section 5.2.3, $\eta^2(S, A)$ is equal to the contribution of A and clustering S to the data scatter. In the case of A_t, analogous roles are played by coefficients G^2 and $X^2 = Q^2$.

Relations between η^2, G^2 and X^2 can be quite complex depending on the bivariate distribution of A and S. However, when the distribution is organized in such a way that all the within-class variances of A are smaller than its overall variance, the pattern of association expressed in G^2 and X^2 generally follows that expressed in η^2.

To illustrate this, let us set an experiment according to the data in Table 2.10: within each of four classes, S_1, S_2, S_3, and S_4, a prespecified number of observations is randomly generated with pre-specified mean and variance. The totality of 2300 generated observations constitutes the quantitative feature A for which the correlation ratio $\eta^2(S, A)$ is calculated. Then, the range of A is divided in $t = 5$ equally-spaced intervals (i.e., not necessarily intervals with an equal number of data) constituting categories of the corresponding attribute A_t, which is cross-classified with S to calculate G^2 and X^2. This setting follows that described in [94].

The initial within-class means are not much different with respect to the corresponding variances. Multiplying each of the initial means by the same factor value, $f = 1, 2, ..., 20$, the means are step by step diverged in such a way that the within-class samples become more and more distinguishable from each other, thus increasing the association between S and A. The final means in Table 2.10 correspond to $f = 20$.

This is reflected in Figure 2.3 where the horizontal axis corresponds to the divergence factor, f, and the vertical axis represents values of the three coefficients for the case when the within class distribution of A is uniform (on the left) or Gaussian, or normal (on the right). We can see that the patterns follow each other rather closely in the case of a uniform distribution. There are small diversions from this in the case of a normal distribution. The product-moment correlation between G^2 and X^2 is always about 0.98-0.99 whereas they both

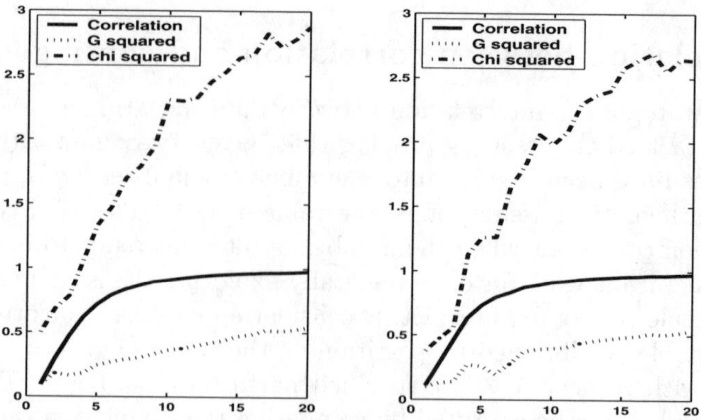

Figure 2.3: Typical change of the correlation ratio (solid line), G squared (dotted line) and chi-square (dashdotted line) with increase of the class divergence factor in the case of uniform (left) and normal (right) within class distribution of the quantitative variable A.

Table 2.10: Setting of the experiment.

Class	S_1	S_2	S_3	S_4
Number of observations	200	100	1000	1000
Variance	1.0	1.0	4.0	0.1
Initial mean	0.5	1.0	1.5	2.0
Final mean	10	20	30	40

correlate with η^2 on the level of 0.90. The difference in values of G^2, X^2 and η^2 is caused by two factors: first, by the coarse qualitative nature of A_t versus the fine-grained quantitative character of A, and, second, by the difference in their contributions to the data scatter. The second factor scales G^2 down and X^2 up, to the maximum value 3 according to (2.15).

2.2.5 Meaning of correlation

Correlation is a phenomenon which may be observed between two features co-occurring in the same observations: the features are co-related in such a way that change in one of them accords with a corresponding change in the other.

These are frequently asked questions: Given a high correlation or association, is there any causal relation behind? Given a low correlation, are the features involved independent? If there is a causal relation, should it translate into a higher correlation?

The answer to each is: no, not necessarily.

To make our point less formal, let us refer to a typical statistics news nugget brought to life by newspapers and BBC Ceefax 25 June 2004: "Children whose

Table 2.11: Association between mother's fish eating (A) and her baby's language skills (B) and health (C).

Feature	B	$\bar{B}$	C	$\bar{C}$	Total
A	520	280	560	240	800
$\bar{A}$	480	720	840	360	1200
Total	1000	1000	1400	600	2000

mothers eat fish regularly during pregnancy develop better language and communication skills. The findings are based on analysis of eating habits of 7400 mothers ... by the University of North Carolina, published in the journal Epidemiology."

At face value, the claim is simple: eat more fish while pregnant and your baby will be better off in the contest of language and communication skills. The real value behind it is a cross classification of a mother's eating habits and her baby's skills over the set of 7400 mother-baby couples at which the cell combining "regular fish eating" with "better language skills" has accounted for a considerably greater number of observations than it would be expected under statistical independence, that is, the corresponding Quetelet coefficient q is positive. So what? Could it be just because of the fish? Very possibly: some say that the phosphorus which is abundant in fish is a building material for the brain. Yet some say that the phosphorus diet has nothing to do with brain development. They think that the correlation is just a manifestation of a deeper relation between family income, not accounted for in the data, and the two features: in richer families it is both the case that mothers eat more expensive food, fish included, and babies have better language skills. The conclusion: more research is needed to see which of these two explanations is correct. And more research may bring further unaccounted for and unforeseen factors and observations.

Example 2.3. False correlation and independence

To illustrate the emergence of "false" correlations and non-correlations, let us dwell on the mother-baby example above involving the following binary features: A – "more fish eating mother," B – "baby's better language skills," and C – "healthier baby." Table 2.11 presents artificial data on two thousand mother-baby couples relating A with B and C.

According to Table 2.11, the baby's language skills (B) are indeed positively related to mother's fish eating (A): 520 observations at cell AB rather than 400 expected if A and B were independent, which is supported by a positive Quetelet coefficient $q(B/A) = 30\%$. In contrast, no relation is observed between fish eating (A) and a baby's health (C): all A/C cross classifying entries on the right of Table 2.11 are proportional to the products of marginal frequencies. For instance, with $p(A) = 0.4$ and $p(C) = 0.7$ their product $p(A)p(C) = 0.28$ accounts for 28% of 2000 observations,

Table 2.12: Association between mother's fish eating (A) and baby's language skills (B) and health (C) with income (D) taken into account.

Feature D	Feature A	B	$\bar{B}$	C	$\bar{C}$	Total
D	A	480	120	520	80	600
	$\bar{A}$	320	80	300	100	400
	Total	800	200	820	180	1000
	A	40	160	40	160	200
$\bar{D}$	$\bar{A}$	160	640	540	260	800
	Total	200	800	580	420	1000
Total		1000	1000	1400	600	2000

that is, 560, which is exactly the entry at cell AC.

However, if we take into account one more binary feature, D, which assigns Yes to better off families, and break down the sample according to D, the data may show a different picture (see Table 2.12). All turned upside down in Table 2.12: what was independent in Table 2.11, A and C, became associated within both D and not-D categories, and what was correlated in Table 2.11, A and B, became independent within both D and not-D categories!

Specifically, with these artificial data, one can see that A accounts for 600 within D category and 200 within not-D category. Similarly, B accounts for 800 within D and only 200 within not-D. Independence between A and B within either strata brings the numbers of AB to 480 in D and only 40 in not-D. This way, the mutually independent A and B within each stratum become correlated in the combined sample, because both A and B are concentrated mostly within D.

Similar though opposite effects are at play with association between A and C: they are negatively related in not-D and positively related in D, so that combining these two strata brings the mutual dependence to zero. □

A high correlation/association is just a pointer to the user, researcher or manager alike, to look at what is behind. The data on their own cannot prove any causal relations, especially when no timing is involved, as is the case in all our exemplary problems. A causal relation can be established only with a mechanism explaining the process in question theoretically, to which the data may or may not add credibility.

2.3 Feature space and data scatter

2.3.1 Data matrix

A quantitative data table is usually referred to as a data matrix. Its rows correspond to entities and columns to variables. Moreover, in most clustering computations, all metadata are left aside so that a feature and entity are represented by the corresponding column and row only, under the assumption that the labels of entities and variables do not change.

A data table with mixed scales such as Table 2.1 will be transformed to a quantitative format. According to rules described in the next section, this is achieved by pre-processing each of the categories into a dummy variable by assigning 1 to an entity that falls in it and 0 otherwise.

Example 2.4. Pre-processing Masterpieces data

Let us convert the Masterpieces data in Table 2.1 to the quantitative format. The binary feature SCon is converted by substituting Yes by 1 and No by zero. A somewhat more complex transformation is performed at the three categories of feature Narrative: each is assigned with a corresponding zero/one vector so that the original column Narrative is converted into three (see Table 2.13). □

Table 2.13: Quantitative representation of the Masterpieces data as an 8 × 7 entity-to-attribute matrix.

Entity	LenSent	LenDial	NumCh	SCon	Objective	Personal	Direct
1	19.0	43.7	2	0	1	0	0
2	29.4	36.0	3	0	1	0	0
3	23.9	38.0	3	0	0	1	0
4	18.4	27.9	2	1	1	0	0
5	25.7	22.3	3	1	0	1	0
6	12.1	16.9	2	1	0	1	0
7	23.9	30.2	4	1	0	0	1
8	27.2	58.0	5	1	0	0	1

A data matrix row corresponding to an entity $i \in I$ constitutes what is called an *M-dimensional point* or *vector* $y_i = (y_{i1}, ..., y_{iM})$ whose components are the row entries. For instance, Masterpieces data in Table 2.13 is a 8 × 7 matrix, and the first row in it constitutes vector $y_1 = (19.0, 43.7, 2, 0, 0, 1, 0)$ each component of which corresponds to a specific feature and, thus, cannot change its position without changing the feature's position in the feature list.

Similarly, a data matrix column corresponds to a feature or category with its elements corresponding to different entities. This is an N-dimensional vector.

Matrix and vector terminology is not just fancy language but part of a well developed mathematical discipline of linear algebra, which is used throughout all data mining disciplines. Some of it will be used in Chapter 5.

2.3.2 Feature space: distance and inner product

Any M-dimensional vector $y = (y_1, ..., y_M)$ pertains to the corresponding combination of feature values. Thus, the set of all M-dimensional vectors y is referred to as the feature space. This space is provided with interrelated distance and similarity measures.

The distance between two M-dimensional vectors, $x = (x_1, x_2, ..., x_M)$ and

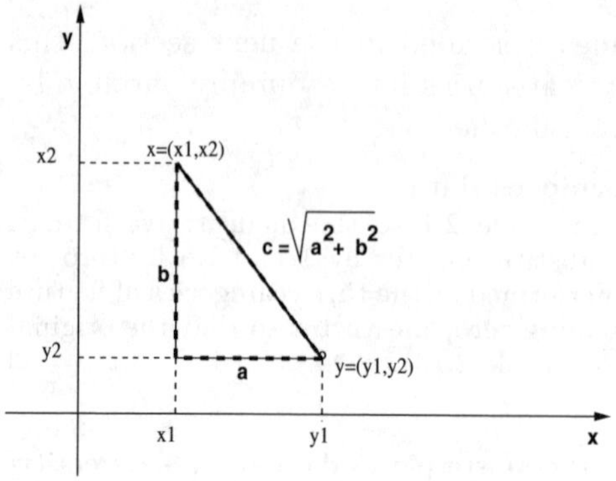

Figure 2.4: Interval between points x and y is the hypotenuse of the highlighted triangle, which explains the distance between x and y.

$y = (y_1, y_2, ..., y_M)$, will be defined as the sum of the component-wise differences squared:

$$d(x, y) = (x_1 - y_1)^2 + (x_2 - y_2)^2 + ... + (x_M - y_M)^2 \tag{2.17}$$

This difference-based quadratic measure is what mathematicians call Euclidean distance squared. It generalizes the basic property of plane geometry, the so-called Pythagoras' theorem as presented in Figure 2.4. Indeed, distance $d(x, y)$ in it is c^2 and c is the Euclidean distance between x and y.

Example 2.5. Distance between entities

Let us consider three novels, 1 and 2 by Dickens, and one, 7 by Tolstoy, as row-points of the matrix in Table 2.13 as presented in the upper half of Table 2.14. The mutual distances between them are calculated in the lower half. The differences in the first two variables, LenS and LenD, predefine the result however different the other features are, because their scales prevail. This way we get a counter-intuitive conclusion that a novel by Dickens is closer to that of Tolstoy than to the other by Dickens because $d(2, 7) = 67.89 < d(1, 2) = 168.45$. Therefore, feature scales must be rescaled to give greater weights to the other variables.

□

The concept of M-dimensional feature space comprises not only all M-series of reals, $y = (y_1, ..., y_M)$, but also two mathematical operations with them: the component-wise summation defined by the rule $x+y = (x_1+y_1, ..., x_n+y_n)$ and multiplication of a vector by a number defined as $\lambda x = (\lambda x_1, ..., \lambda x_n)$ for any real λ. These operations naturally generalize the operations with real numbers and have similar properties.

With such operations, variance s_v^2 also can be expressed as the distance, between column $v \in V$ and column vector $c_v e$, whose components are all equal to the column's average, c_v, divided by N, $s_v^2 = d(x_v, c_v e)/N$. Vector $c_v e$ is the

Table 2.14: Computation of distances between three masterpieces according to Table 2.13. Squared differences of values in the upper part are in the lower part of the matrix, column-wise; they are summed up in the column "Distance" on the right.

Item	LenS	LenD	NumC	SCon	Obje	Pers	Dire	Distance
1	19.0	43.7	2	0	1	0	0	
2	29.4	36.0	3	0	1	0	0	
7	23.9	30.2	4	1	0	0	1	
Distance								Total
d(1,2)	108.16	59.29	1	0	0	0	0	168.45
d(1,7)	24.01	182.25	4	1	1	0	1	213.26
d(2,7)	30.25	33.64	1	1	1	0	1	67.89

result of multiplication of vector $e = (1, ..., 1)$ whose components are all unities by c_v.

Another important operation is the so-called inner, or scalar, product. For any two M-dimensional vectors x, y their inner product is a number denoted by (x, y) and defined as the sum of component-wise products, $(x, y) = x_1y_1 + x_2y_2 + ... + x_My_M$.

The inner product and distance are closely related. It is not difficult to see, just from the definitions, that for any vectors/points x, y: $d(x, 0) = (x, x) = \sum_{v \in V} x_v^2$ and $d(y, 0) = (y, y)$ and, moreover, $d(x, y) = (x-y, x-y)$. The symbol 0 refers here to a vector with all components equal to zero. The distance $d(y, 0)$ will be referred to as the scatter of y. The square root of the scatter $d(y, 0)$ is referred to as the (Euclidean) norm of y and denoted by $||y|| = \sqrt{(y, y)} = \sqrt{\sum_{i \in I} y_i^2}$. It expresses the length of y.

In general, for any M-dimensional x, y, the following equation holds:

$$d(x, y) = (x - y, x - y) = (x, x) + (y, y) - 2(x, y) = d(0, x) + d(0, y) - 2(x, y). \tag{2.18}$$

This equation becomes especially simple when $(x, y) = 0$. In this case, vectors x, y are referred to as mutually *orthogonal*. When x and y are mutually orthogonal, $d(0, x - y) = d(0, x + y) = d(0, x) + d(0, y)$, that is, the scatters of $x - y$ and $x + y$ are equal to each other and the sum of scatters of x and y. This is a multidimensional analogue to the Pythagoras theorem and the base for decompositions of the data scatter employed in many statistical theories including the theory for clustering presented in Chapter 5.

The inner product of two vectors has a simple geometric interpretation (see Figure 2.1 on page 48), $(x, y) = ||x||||y|| \cos \alpha$ where α is the "angle" between x and y (at 0). This conforms to the concept of orthogonality above: vectors are orthogonal when the angle between them is a right angle.

2.3.3 Data scatter

The summary scatter of all row-vectors in data matrix Y is referred to as the data scatter of Y and denoted by

$$T(Y) = \sum_{i \in I} d(0, y_i) = \sum_{i \in I} \sum_{v \in V} y_{iv}^2 \tag{2.19}$$

Equation (2.19) means that $T(Y)$ is the total of all Y entries squared.

An important characteristic of feature $v \in V$ is its *contribution to the data scatter* defined as

$$T_v = \sum_{i \in I} y_{iv}^2, \tag{2.20}$$

the distance of the N-dimensional column from the zero column. Data scatter is obviously the sum of contributions of all variables, $T(Y) = \sum_{v \in V} T_v$.

If feature v is centered, then its contribution to the data scatter is proportional to its variance:

$$T_v = N s_v^2 \tag{2.21}$$

Indeed, $c_v = 0$ since v is centered. Thus, $s_v^2 = \sum_{i \in I} (y_{iv} - 0)^2 / N = T_v / N$.

The relative contribution $T_v / T(Y)$ is a characteristic playing an important role in data standardization issues as explained in the next section.

2.4 Pre-processing and standardizing mixed data

The data pre-processing stage is to transform the raw entity-to-feature table into a quantitative matrix for further analysis. To do this, one needs first to convert all categorical data to a numerical format. We will do this by using a dummy zero-one variable for each category. Then variables are standardized by shifting their origins and rescaling. This operation can be clearly substantiated from a statistics perspective, typically, by assuming that entities have been randomly sampled from an underlying Gaussian distribution. In data mining, substantiation may come from the data geometry. By shifting all the origins to feature means, entities become scattered around the center of gravity so that clusters can be more easily "seen" from that point. With feature rescaling, feature scales become balanced according to the principle of equal importance of each feature brought into the data table.

To implement these general principles, we are going to rely on the following three-stage procedure. The stages are: (1) enveloping qualitative categories, (2) standardization, and (3) rescaling, as follows:

1. **Quantitatively enveloping categories**: This stage is to convert a mixed scale data table into a quantitative matrix by treating every qualitative category as a separate dummy variable coded by 1 or 0 depending on whether an entity falls into the category or not. Binary features are coded similarly except that no additional columns are created. Quantitative features are left as they are. The converted data table will be denoted by $X = (x_{iv})$, $i \in I, v \in V$.

2. **Standardization**: This stage aims at transforming feature-columns of the data matrix to make them comparable by shifting their origins to a_v and rescaling them by b_v, $v \in V$, thus to create standardized matrix $Y = (y_{iv})$:

$$y_{iv} = \frac{x_{iv} - a_v}{b_v}, \; i \in I, v \in V. \tag{2.22}$$

   In this text, the shift coefficient a_v always will be the grand mean. In particular, the dummy variable corresponding to category $v \in V_l$ has its mean $c_v = p_v$, the proportion of entities falling in the category.

   The scale factor b_v can be either the standard deviation or range or other quantity reflecting the variable's spread. In particular, for a category $v \in V_l$, the standard deviation can be either $\sqrt{p_v(1-p_v)}$ (Bernoulli distribution) or $\sqrt{p_v}$ (Poisson distribution), see page 46. The range of a dummy variable is always 1.

   Using the standard deviation is popular in data mining probably because it is used in classical statistics relying on the theory of Gaussian distribution which is characterized by the mean and standard deviation. Thus standardized, contributions of all features to data scatter become equal to each other because of the proportionality of contributions and standard deviations. On first glance this seems an attractive property guaranteeing equal contributions of all features to the results, an opinion to which the current author once also subscribed [90]. However, this is not so. Two different factors contribute to the value of standard deviation: the feature scale and the shape of its distribution. As shown in section 2.1.2, within the same range scale the standard deviation may greatly vary from the minimum, at the peak unimodal distribution, to the maximum, at the peak bimodal distribution. By standardizing with standard deviations, we deliberately bias data in favor of unimodal distributions, although obviously it is the bimodal distribution that should contribute to clustering most. This is why the range, not the standard deviation, is used here as the scaling factor b_v. In the case when there can be outliers in data, which may highly affect the range, another more stable range-like scaling factor can be chosen, such as the difference between percentiles, that does not

Table 2.15: Std standardized Masterpieces matrix; Mean is grand mean, Std the standard deviation and Cntr the relative contribution of a variable.

	LS	LD	SC	NC	Ob	Pe	Di
1	-0.6	0.7	-0.9	-1.2	1.2	-0.7	-0.5
2	1.2	0.1	0.0	-1.2	1.2	-0.7	-0.5
3	0.3	0.3	0.0	-1.2	-0.7	1.2	-0.5
4	-0.7	-0.5	-0.9	0.7	1.2	-0.7	-0.5
5	0.6	-0.9	0.0	0.7	-0.7	1.2	-0.5
6	-1.8	-1.3	-0.9	0.7	-0.7	1.2	-0.5
7	0.3	-0.3	0.9	0.7	-0.7	-0.7	1.6
8	0.8	1.8	1.9	0.7	-0.7	-0.7	1.6
Mean	22.4	34.1	3.0	0.6	0.4	0.4	0.3
Std	5.6	12.9	1.1	0.5	0.5	0.5	0.5
Cntr, %	14.3	14.3	14.3	14.3	14.3	14.3	14.3

much depend on the distribution shape. The range based scaling option has been supported experimentally in [87].

3. **Rescaling**: This stage rescales column-features v, which come from the same categorical variable l, back by further dividing y_{iv} with supplementary rescaling coefficients b'_v to restore the original weighting of raw variables. The major assumption in clustering is that all raw variables are supposed to be of equal weight. Having its categories enveloped, the "weight" of a nominal variable l becomes equal to the summary contribution $T_l = \sum_{v \in V_l} T_v$ to the data scatter where V_l is the set of categories belonging to l. Therefore, to restore the original "equal weighting" of l, the total contribution T_l must be related to $|V_l|$, which is achieved by taking $b'_v = \sqrt{|V_l|}$ for $v \in V_l$.

   For a quantitative $v \in V$, b'_v is, typically, unity.

   Sometimes, there can be available an expert evaluation of the relative weights of the original variables l. If such is the case, rescaling coefficients b'_v should be redefined with the square roots of the expert supplied relative weights. This option may be applied to both quantitative and qualitative features.

Note that two of the three steps above refer to categorical features.

Example 2.6. Effects of different scaling options

Table 2.15 presents the Masterpieces data in Table 2.13 standardized according to the most popular transformation of feature scales, the so-called z-scoring, when the scales are shifted to their mean values and then normalized by the standard deviations. Table 2.16 presents the same data matrix range standardized. All feature contributions are different in this table except for those of NC, Ob and Pe which are the same. Why the same? Because they have the same variance $p(1-p)$ corresponding to p or $1-p$ equal to 3/8.

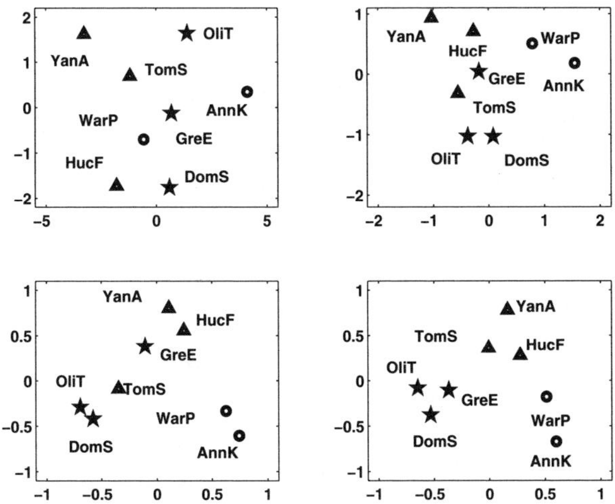

Figure 2.5: Masterpieces on the plane of two first principal components at four different standardizations: no scaling (top left), scaling with standard deviations (top right), range scaling (bottom left), and range scaling with the follow-up rescaling (bottom right).

We can see how overrated the summary contribution of the qualitative variable Narrative becomes: three dummy columns on the right in Table 2.16 take into account 55.7% of the data scatter and thus highly affect any further results. This is why further rescaling of these three variables by the $\sqrt{3}$ is needed to decrease their total contribution 3 times. Table 2.17 presents results of this operation applied to data in Table 2.16.

Note that the total Narrative's contribution per cent has not changed as much as we would expect: about two times, yet not three times!

Figure 2.5 shows how important the scaling can be for clustering results. It displays mutual locations of the eight masterpieces on the plane of the first two principal components of the data in Table 2.13 at different scaling factors: (a) left top: no scaling at all; (b) right top: scaling by the standard deviations, see Table 2.15; (c) left bottom: scaling by ranges; (d) right bottom: scaling by ranges with the follow-up rescaling of the three dummy variables for categories of Narrative by taking into account that they come from the same nominal feature; the scale shifting parameter is always the variable's mean.

The left top scatter-plot displays no relation to the novels' authorship. Probably no clustering algorithm can properly identify the author classes with this standardization of the data (Table 2.15). On the contrary, the authorship pattern is clearly displayed on the bottom right figure, and it is likely that any reasonable clustering algorithm will capture them with this standardization.

We can clearly see on Figure 2.5 that, in spite of the unidimensional nature of transformation (2.22), its combination of shifts and scales can be quite powerful in changing the geometry of the data. □

Table 2.16: Range standardized Masterpieces matrix; Mean is grand mean, Range the range and Cntr the relative contribution of a variable.

	LS	LD	SC	NC	Ob	Pe	Di
1	-0.2	0.2	-0.3	-0.6	0.6	-0.4	-0.3
2	0.4	0.0	0.0	-0.6	0.6	-0.4	-0.3
3	0.1	0.1	0.0	-0.6	-0.4	0.6	-0.3
4	-0.2	-0.2	-0.3	0.4	0.6	-0.4	-0.3
5	0.2	-0.3	0.0	0.4	-0.4	0.6	-0.3
6	-0.6	-0.4	-0.3	0.4	-0.4	0.6	-0.3
7	0.1	-0.1	0.3	0.4	-0.4	-0.4	0.8
8	0.3	0.6	0.7	0.4	-0.4	-0.4	0.8
Mean	22.4	34.1	3.0	0.6	0.4	0.4	0.3
Range	17.3	41.1	3.0	1.0	1.0	1.0	1.0
Cntr, %	7.8	7.3	9.4	19.9	19.9	19.9	15.9

Table 2.17: Range standardized Masterpieces matrix with the additionally rescaled nominal feature attributes; Mean is grand mean, Range the range and Cntr the relative contribution of a variable.

	LS	LD	SC	NC	Ob	Pe	Di
1	-0.20	0.23	-0.33	-0.63	0.36	-0.22	-0.14
2	0.40	0.05	0.00	-0.63	0.36	-0.22	-0.14
3	0.08	0.09	0.00	-0.63	-0.22	0.36	-0.14
4	-0.23	-0.15	-0.33	0.38	0.36	-0.22	-0.14
5	0.19	-0.29	0.00	0.38	-0.22	0.36	-0.14
6	-0.60	-0.42	-0.33	0.38	-0.22	0.36	-0.14
7	0.08	-0.10	0.33	0.38	-0.22	-0.22	0.43
8	0.27	0.58	0.67	0.38	-0.22	-0.22	0.43
Mean	22.45	34.13	3.00	0.63	0.38	0.38	0.25
Range	17.30	41.10	3.00	1.00	1.73	1.73	1.73
Cntr, %	12.42	11.66	14.95	31.54	10.51	10.51	8.41

Example 2.7. Relative feature weighting under standard deviations and ranges may differ

For the Market town data, with $N = 45$ and $n = 12$, the summary feature characteristics are shown in Table 2.18.

As proven above, the standard deviations in Table 2.18 are at least twice as small as the ranges, which is true for all data tables. However, the ratio of the range over the standard deviation may differ for different features reaching as much as 3/0.6=5 for DIY. Therefore, using standard deviations and ranges for scaling in (2.22) may lead to differences in relative scales between the variables and, thus, to different clustering results as well as in Masterpiece data. □

Are there any regularities in the effects of data standardization (and rescaling) on the data scatter and feature contributions to it? Not many. But there are items that should be mentioned:

Table 2.18: Summary characteristics of the Market town data; Mean is grand mean, Std the standard deviation, Range the range and Cntr the relative contribution of a variable.

	P	PS	Do	Ho	Ba	Su	Pe	DIY	SP	PO	CAB	FM
Mean	7351.4	3.0	1.4	0.4	4.3	1.9	2.0	0.2	0.5	2.6	0.6	0.2
Std	6193.2	2.7	1.3	0.6	4.4	1.7	1.6	0.6	0.6	2.1	0.6	0.4
Range	21761.0	12.0	4.0	2.0	19.0	7.0	7.0	3.0	2.0	8.0	2.0	1.0
Cntr, %	8.5	5.4	11.1	8.8	5.6	6.3	5.7	4.2	10.3	7.3	9.7	17.1

Effect of shifting to the average. With shifting to the averages, the feature contributions and variances become proportional to each other, $T_v = Ns_v^2$, for all $v \in V$.

Effects of scaling of categories. Let us take a look at the effect of scaling and rescaling coefficients on categories. The contribution of a binary attribute v, standardized according to (2.22), becomes $T_v = Np_v(1 - p_v)/(b_v)^2$ where p_v is the relative frequency of category v. This can be either $p_v(1-p_v)$ or 1 or $1-p_v$ depending on whether $b_v = 1$ (range) or $b_v = \sqrt{p_v(1-p_v)}$ (Bernoulli standard deviation) or $b_v = \sqrt{p_v}$ (Poisson standard deviation), respectively. These can give some guidance in rescaling of binary categories: the first option should be taken when both zeros and ones are equally important, the second when the distribution does not matter and the third when it is only unities that matter.

Identity of binary and two-category features. An important issue faced by the user is how to treat a categorical feature with two categories such as gender (Male/Female) or voting (Democrat/Republican) or belongingness to a group (Yes/No). The three-step procedure of standardization makes the issue irrelevant: there is no difference between a two-category feature and either of its binary representations.

Indeed, let x be a two-category feature assigning each entity $i \in I$ with a category 'eks1' or 'eks2' whose relative frequencies are p_1 and p_2 such that $p_1+p_2 = 1$. Denote by $y1$ a binary feature corresponding to category 'eks1' so that $y1_i = 1$ if $x_i =$ eks1 and $y1_i = 0$ if $x_i =$ eks2. Analogously define a binary feature $y2$ corresponding to category 'eks2'. Obviously, $y1$ and $y2$ complement each other so that their sum makes an all unity vector.

In the first stage of the standardization procedure, the user decides whether x is to be converted to a binary feature or left as a categorical one. In the former case, x is converted to a dummy feature, say, column $y1$; and in the latter case, x is converted into a two-column submatrix consisting of columns $y1$ and $y2$. Since the averages of $y1$ and $y2$ are p_1 and p_2, respectively, after shifting column $y1$'s entries become $1-p_1$, for 1, and $-p_1$, for 0. Respective entries of $y2$ are shifted to $-p2$ and $1-p2$, which can be expressed through $p_1 = 1-p_2$ as $p_1 - 1$ and p_1. That means that $y_2 = -y_1$ after the shifting: the two columns become identical, up to the sign. That implies that all their square contribution characteristics become the same, including the total contributions to the data scatter so that the total contribution of the two-column submatrix is twice greater than the contribution of a single column $y1$, whichever scaling option is accepted. However, further rescaling the two-column submatrix by the recommended $\sqrt{2}$ restores the balance of contributions: the two-column submatrix contributes as much as a single column.

Total contributions of categorical features. The total contribution of nominal variable l is

$$T_l = N \sum_{v \in V_l} p_v(1-p_v)/(b_v)^2. \tag{2.23}$$

Depending on the choice of scaling coefficients b_v, this can be

1. $T_l = N(1 - \sum_{v \in V_l} p_v^2)$ if $b_v = 1$ (range normalization);
2. $T_l = N|V_l|$ if $b_v = \sqrt{p_v(1-p_v)}$ (Bernoulli normalization);
3. $T_l = N(|V_l| - 1)$ if $b_v = \sqrt{p_v}$ (Poisson normalization).

where $|V_l|$ is the number of categories of l. The quantity on the top is the Gini coefficient of the distribution of v (2.3).

The square roots of these should be used for further rescaling qualitative categories stemming from the same nominal variable l to adjust their total impact on the data scatter.

2.5 Other table data types

2.5.1 Dissimilarity and similarity data

In many cases the entity-to-entity dissimilarity or similarity data is the preferred format of data as derived from more complex data such as Primates in Table 1.2 or as directly resulting from observations as Confusion data in Table 1.9.

Similarity scoring is especially important for treating the so-called "wide" data tables in which the number of features is much greater than the number of entities. Such is the case of unstructured textual documents for which the presence or absence of a keyword is a feature. The number of meaningful keywords may go into hundreds of thousands even when the entire text collection is in dozens or hundreds. In this case, bringing in a text-to-text similarity index may convert the problem from a virtually untreatable one into a modest size clustering exercise.

An entity-to-entity similarity index may appear as the only data for clustering. For example, similarity scores may come from experiments on subjective judgements such as scoring individual's evaluation of similarity between stimuli or products. More frequently, though, similarities are used when entities to be clustered are too complex to be put in the entity-to-feature table format. When considering two biomolecular amino acid sequences (proteins) a similarity score between them can be based on the probability of transformation of one of them into the other with evolutionary meaningful operations of substitution, deletion and insertion of amino acids [70].

The terminology reflects the differences between the two types of proximity scoring: the smaller the dissimilarity coefficient the closer the entities are, whereas the opposite holds for similarities. Also, the dissimilarity is conventionally considered as a kind of extended distance, thus satisfying some distance properties. In particular, given a matrix $D = (d_{ij})$, $i, j \in I$, where I is the entity set, the entries d_{ij} form a dissimilarity measure between entities $i, j \in I$ if D satisfies the properties:

(a) Symmetry: $d_{ij} = d_{ji}$.

(b) Non-negativity: $d_{ij} \geq 0$.

(c) Semi-definiteness: $d_{ij} = 0$ if entities i and j coincide.

A dissimilarity measure is referred to as a distance if it additionally satisfies:

(d) Definiteness: $d_{ij} = 0$ if and only if entities i and j coincide.

(e) Triangle inequality: $d_{ij} \leq d_{il} + d_{lj}$ for any $i, j, l \in I$.

A distance is referred to as an ultra-metric if it satisfies a stronger triangle inequality:

(f) Ultra-triangle inequality: $d_{ij} \leq \max(d_{il}, d_{lj})$ for any $i, j, l \in I$.

In fact, the ultra-triangle inequality states that among any three distances d_{ij}, d_{il} and d_{jl}, two are equal to each other and the third cannot be greater than that. Ultra-metrics emerge as distances between leaves of trees; in fact, they are equivalent to some tree structures such as the heighted upper cluster hierarchies considered in section 5.3.1.

No such properties are assumed for similarity data except, sometimes, for symmetry (a).

Standardization of similarity data

Given a similarity measure, all entity-to-entity similarities are measured in the same scale so that its change will not change clustering results. This is why there is no need to change the scale of similarity data. As to the shift in the origin of the similarity measure, this can be of advantage by making within- and between-cluster similarities more contrasted. Figure 2.6 demonstrates the effect

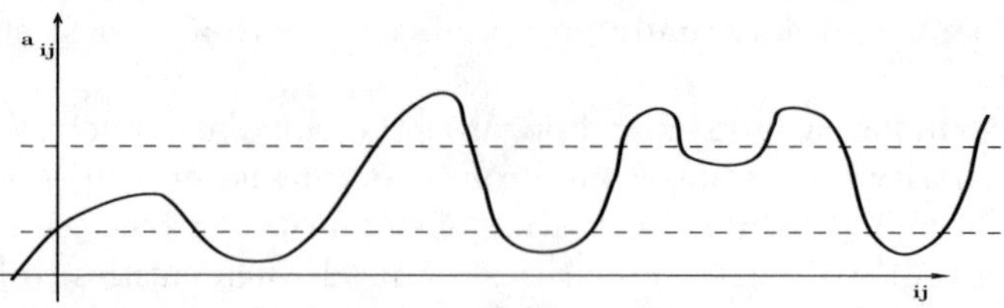

Figure 2.6: A pattern of similarity $a_{ij} = s_{ij} - a$ values depending on a subtracted threshold a.

of changing a positive similarity measure s_{ij} to $a_{ij} = s_{ij} - a$ by subtracting a threshold $a > 0$: small similarities $s_{ij} < a$ can be transformed into negative similarities a_{ij}. This can be irrelevant as for example in such clustering methods as single linkage or similarity-based K-Means. But there are methods such as ADDI-S in section 5.5.5 which can be quite sensitive to the threshold a.

Shift of the origin can be a useful option in standardizing similarity data.

Standardization of dissimilarity data

Given a dissimilarity measure d_{ij}, $i, j \in I$, it is frequently standardized by transforming it into both a row- and column-wise centered similarity measure according to the formula:

$$s_{ij} = -(d_{ij} - d_{i\cdot} - d_{\cdot j} - d_{\cdot\cdot})/2 \tag{2.24}$$

where the dot denotes the operation of averaging so that $d_{i\cdot} = \sum_{j\in I} d_{ij}/N$, $d_{\cdot j} = \sum_{i\in I} d_{ij}/N$, and $d_{\cdot\cdot} = \sum_{i,j\in I} d_{ij}/(N \times N)$.

This formula can be applied to any dissimilarity measure, but it is especially suitable in the situation in which d_{ij} is Euclidean distance squared, that is, when $d_{ij} = (x_i - x_j, x_i - x_j)$ for some multidimensional x_i, x_j, $i, j \in I$. It can be proven then that s_{ij} in (2.24) is the inner product $s_{ij} = (x_i, x_j)$ for all $i, j \in I$ if all x_i are centered.

2.5.2 Contingency and flow data

Table $F = (f_{ij})$, $i \in I$, $j \in J$, is referred to as a flow table if every entry expresses a quantity of the same matter in such a way that all of the entries can be meaningfully summed up to a number expressing the total amount of the matter in the data. Examples of flow data tables are: (a) contingency tables counting numbers of co-occurred instances; (b) mobility tables counting numbers of individual members of a group having changed their categories; (c) trade tables showing the money transferred from i to j during a specified period.

This type of data is of particular interest in processing massive information sources. The data itself can be untreatable within time-memory constraints, but by counting co-occurrences of categories of interest in a sample it can be pre-processed into the flow data format and analyzed as such.

The nature of the data associates weights of row categories $i \in I$, $f_{i+} = \sum_{j \in J} f_{ij}$, and column categories $j \in J$, $f_{+j} = \sum_{i \in I} f_{ij}$, the total flow from i and that to j. The total flow volume is $f_{++} = \sum_{i \in I} \sum_{j \in J} f_{ij}$, which is the summary weight, $f_{++} = \sum_{j \in J} f_{+j} = \sum_{i \in I} f_{i+}$. Extending concepts introduced in section 2.2.3 for contingency tables to the general flow data, we can extend the definition of the relative Quetelet index as

$$q_{ij} = \frac{f_{ij} f_{++}}{f_{i+} f_{+j}} - 1 \tag{2.25}$$

This index, in fact, compares the share of j in i's transaction, $p(j/i) = f_{ij}/f_{i+}$, with the share of j in the overall flow, $p(j) = f_{+j}/f_{++}$.

Indeed, it is easy to see that q_{ij} is the relative difference of the two, $q_{ij} = (p(j/i) - p(j))/p(j)$. Obviously, $q_{ij} = 0$ when there is no difference.

Transformation (2.25) is a standardization of flow data which takes into account the data's nature. Standardization (2.25) is very close to the so-called normalization of Rao and Sabovala widely used in marketing research and, also, the (marginal) cross-product ratio utilized in the analysis of contingency data. Both can be expressed as p_{ij} transformed into $q_{ij} + 1$.

Example 2.8. Quetelet coefficients for Confusion data Coefficients (2.25) are presented in Table 2.19. One can see from the table that the numerals overwhelmingly respond to themselves. However, there are also a few positive entries in the table outside of the diagonal. For instance, 7 is perceived as 1 with the frequency 87.9% greater than the average, and 3 is perceived as 9, and vice versa, with also higher frequencies than the average. □

Table 2.19: Relative Quetelet coefficients for Confusion data, per cent.

Sti-	Response									
mulus	1	2	3	4	5	6	7	8	9	0
1	512.7	-92.6	-92.7	-75.2	-95.8	-83.0	-31.8	-100.0	-95.9	- 96.5
2	-90.2	728.9	-50.8	-95.5	-61.7	-46.7	-84.0	-69.6	-92.9	- 84.1
3	-79.7	-69.3	612.1	-92.1	-80.9	-100.0	-54.5	-69.7	54.6	- 86.8
4	4.1	-76.7	-95.8	725.9	-95.8	-87.6	-65.9	-92.7	-58.3	- 100.0
5	-90.2	-72.5	-55.0	-84.2	610.7	-10.6	-92.0	-92.7	28.3	- 87.6
6	-82.5	-85.2	-92.7	-87.6	2.9	615.8	-95.5	61.9	-88.8	- 62.1
7	87.9	-95.8	-78.0	-76.3	-92.6	-100.0	658.6	-100.0	-95.9	- 93.8
8	-92.3	-70.4	-70.7	-79.7	-80.9	-20.8	-87.5	502.7	-31.9	51.6
9	-82.5	-69.3	16.1	-48.1	-13.0	-87.6	-76.1	-14.3	459.3	- 62.1
0	-87.4	-95.8	-92.7	-87.6	-92.6	-79.6	-71.6	-25.8	-78.6	621.1

Taking into account that the flow data entries can be meaningfully summed up to the total flow volume, the distance between row entities in a contingency table is defined by weighting the columns by their "masses" p_{+l} as follows,

$$\chi(k, k') = \sum_{j \in J} p_{+j}(q_{kj} - q_{k'j})^2. \tag{2.26}$$

This is equal to the so-called chi-square distance defined between row conditional profiles in the Correspondence factor analysis (see section 5.1.4). The formula (2.26) is similar to formula (2.17) for Euclidean distance squared except for the weights. A similar chi-square distance can be defined for columns.

The concept of data scatter for contingency data is also introduced with weighting:

$$X^2(P) = \sum_{I \in I} \sum_{j \in J} p_{i+} p_{+j} q_{ij}^2 \tag{2.27}$$

The notation reflects the fact that this value is closely connected with the Pearson chi-square contingency coefficient, defined in (2.13) above (thus to Q^2 (2.12) as well). Elementary algebraic manipulations show that $X^2(P) = X^2$. Note that in this context the chi-squared coefficient has nothing to do with the statistical independence: it is just a weighted data scatter measure compatible with the specific properties of flow and contingency data. In particular, it is not difficult to prove an analogue to a property of the conventional data scatter: $X^2(P)$ is the sum of chi-square distances $\chi(k, 0)$ over all k.

Thus, Pearson's chi-squared coefficient here emerges as the data scatter after the data have been standardized into Quetelet coefficients. For Table 1.9 transformed here into Table 2.19 the coefficient is equal to 4.21 which is 47% of 9, the maximum value of the coefficient for 10×10 tables.

Chapter 3

K-Means Clustering

After reading this chapter the reader will know about:

1. Straight and incremental K-Means.
2. The instability of K-Means with regard to initial centroids.
3. An anomalous cluster version of K-Means for incomplete clustering.
4. Three approaches to the initial setting in K-Means: random, maxmin and anomalous pattern.
5. An intelligent version of K-Means mitigating issues of the initial setting and interpretation.
6. Cross-validation of clustering results.
7. Conventional and contribution based interpretation aids for K-Means.

Base words

Anomalous pattern A method for separating a cluster which is most distant from the so-called reference point, which may coincide with the grand mean of the entity set. The method works as K-Means at K=2 except for the location of the reference point which is never changed.

Centroid A multidimensional vector minimizing the summary distance to cluster's elements. If the distance is Euclidean squared, the centroid is equal to the center of gravity of the cluster.

Cluster representative An entity that is considered to represent its cluster well. Conventionally such an entity is drawn as that nearest to the cluster centroid in the Euclidean space. The theory used here suggests that the nearest entity must be drawn according to the inner product rather than distance, which extends the cluster tendencies over the grand mean.

Contributions to the data scatter Additive items representing parts of the data scatter that are explained by certain elements of a cluster structure such as feature-cluster pairs. The greater the contribution, the more important the element. Summary contributions coincide with statistical measures of correlation and association, which is a theoretical support to the recommended data standardization rules.

Cross validation A procedure for testing consistency of a clustering algorithm or its results by the comparison of cluster results found on subsamples formed by a random partitioning of the entity set into a number of groups of equal sizes.

iK-Means An intelligent version of K-Means, in which an initial set of centroids (seeds) is found with an iterated version of the Anomalous pattern algorithm.

Incremental K-Means A version of K-Means in which entities are dealt with one-by-one.

Interpretation aids Computational tools for helping the user to interpret clusters in terms of features, external or used in the process of clustering. Conventional interpretation aids include cluster centroids and bivariate distributions of cluster partitions and features. Contribution based interpretation aids such as ScaD and QScaD tables are derived from the decomposition of the data scatter into parts explained and unexplained by the clustering.

K-Means A major clustering method producing a partition of the entity set into non-overlapping clusters along with within-cluster centroids. It proceeds in iterations consisting of two steps each; one step updates clusters according to the Minimum distance rule, the other step updates centroids as the centers of gravity of clusters. The method implements the so-called alternating minimization algorithm for the square error criterion. To initialize the computations, either a partition or a set of all K tentative centroids must be specified.

Minimum distance rule The rule which assigns each of the entities to its nearest centroid.

Reference point A vector in the variable space serving as the space origin. The Anomalous pattern is sought starting from an entity which is the farthest from the reference point, which thus models the norm from which the Anomalous pattern deviates most.

ScaD and QScaD tables Interpretation aids helping to capture cluster-specific features that are relevant to K-Means clustering results. ScaD is a cluster-to-feature table whose entries are cluster-to-feature contributions to the data scatter. QScaD is a table of the relative Quitelet coefficients of the ScaD entries to express how much they differ from the average.

Square error criterion The sum of summary distances from cluster elements to the cluster centroids, which is minimized by K-Means. The distance used is the Euclidean distance squared, which is compatible with the least-squares data recovery criterion.

3.1 Conventional K-Means

3.1.1 Straight K-Means

K-Means is a major clustering technique that is present, in various forms, in major statistical packages such as SPSS [42] and SAS [17, 119] and data mining packages such as Clementine [14], iDA tool [114] and DBMiner [44].

The algorithm is appealing in many aspects. Conceptually it may be considered a model for the human process of making a typology. Also, it has nice mathematical properties. This method is computationally easy, fast and memory-efficient. However, there are some problems too, especially with respect to the initial setting and stability of results, which will be dealt with in section 3.2.

The cluster structure in K-Means is a partition of the entity set in K non-overlapping clusters represented by lists of entities and within cluster means of the variables. The means are aggregate representations of clusters and as such they are sometimes referred to as standard points or centroids or prototypes. These terms are considered synonymous in the remainder of the text. More formally, the cluster structure is represented by subsets $S_k \subset I$ and M-dimensional centroids $c_k = (c_{kv})$, $k = 1, ..., K$. Subsets S_k form partition $S = \{S_1, ..., S_K\}$ with a set of centroids $c = \{c_1, ..., c_K\}$.

Example 3.9. Centroids of author clusters in Masterpieces data

Let us consider the author-based clusters in the Masterpieces data. The cluster structure is presented in Table 3.1 in such a way that the centroids are calculated twice, once for the raw data in Table 2.13 and the second time, with the standardized data in Table 3.2, which is a copy of Table 2.17 of the previous chapter.

□

Given K M-dimensional vectors c_k as cluster centroids, the algorithm updates cluster lists S_k according to the so-called *Minimum distance rule.*

Minimum distance rule assigns entities to their nearest centroids. Specif-

Table 3.1: Means of the variables in Table 3.2 within K=3 author-based clusters, real (upper row) and standardized (lower row).

Cl.	List	Mean						
		LS (f1)	LD (f2)	NC (f3)	SC (f4)	P (f5)	O (f6)	D (f7)
1	1, 2, 3	24.1	39.2	2.67	0	0.67	0.33	0
		0.095	0.124	-0.111	-0.625	0.168	-0.024	-0.144
2	4, 5, 6	18.7	22.4	2.33	1	0.33	0.67	0
		-0.215	-0.286	-0.222	0.375	-0.024	0.168	-0.144
3	7, 8	25.6	44.1	4.50	1	0.00	0.00	1
		0.179	0.243	0.500	0.375	-0.216	-0.216	0.433

Table 3.2: Range standardized Masterpieces matrix with the additionally rescaled nominal feature attributes copied from Table 2.17.

	LS	LD	NC	SC	Ob	Pe	Di
1	-0.20	0.23	-0.33	-0.63	0.36	-0.22	-0.14
2	0.40	0.05	0.00	-0.63	0.36	-0.22	-0.14
3	0.08	0.09	0.00	-0.63	-0.22	0.36	-0.14
4	-0.23	-0.15	-0.33	0.38	0.36	-0.22	-0.14
5	0.19	-0.29	0.00	0.38	-0.22	0.36	-0.14
6	-0.60	-0.42	-0.33	0.38	-0.22	0.36	-0.14
7	0.08	-0.10	0.33	0.38	-0.22	-0.22	0.43
8	0.27	0.58	0.67	0.38	-0.22	-0.22	0.43

ically, for each entity $i \in I$, its distances to all centroids are calculated, and the entity is assigned to the nearest centroid. When there are several nearest centroids, the assignment is taken among them arbitrarily. In other words, S_k is made of all such $i \in I$ that $d(i, c_k)$ is minimum over all centroids from $c = \{c_1, ..., c_K\}$. The Minimum distance rule is popular in data analysis and can be found under different names such as Voronoi diagrams and vector learning quatization.

In general, some centroids may be assigned no entity at all with this rule.

Having cluster lists updated with the Minimum distance rule, the algorithm updates centroids as gravity centers of the cluster lists S_k; the gravity center coordinates are defined as within cluster averages, that is, updated centroids are defined as $c_k = c(S_k)$, $k = 1, ..., K$, where $c(S)$ is a vector whose components are averages of features over S.

Then the process is reiterated until clusters do not change.

Recall that the distance referred to is Euclidean squared distance defined, for any M-dimensional $x = (x_v)$ and $y = (y_v)$ as $d(x, y) = (x_1 - y_1)^2 + ... + (x_M - y_M)^2$.

Example 3.10. Minimum distance rule at author cluster centroids in Masterpieces data

Let us apply the Minimum distance rule to entities in Table 3.2, given the standardized centroids in Table 3.1. The matrix of distances between the standardized eight row points in Table 3.2 and three centroids in Table 3.1 is in Table 3.3. The table shows that points 1,2,3 are nearest to centroid c_1, 4,5,6 to c_2, and 7, 8 to c_3, which is boldfaced. This means that the rule does not change clusters. These clusters will have the same centroids. Thus, no further calculations can change the clusters: the author-based partition is to be accepted as the result. □

Let us now explicitly formulate the algorithm, which will be referred to as straight K-Means. Sometimes the same procedure is referred to as batch K-Means or parallel K-Means.

Table 3.3: Distances between the eight standardized Masterpiece entities and centroids; within column minima are highlighted.

Centroid	Entity, row point from Table 3.2							
	1	2	3	4	5	6	7	8
	1	2	3	4	5	6	7	8
c_1	**0.22**	**0.19**	**0.31**	1.31	1.49	2.12	1.76	2.36
c_2	1.58	1.84	1.36	**0.33**	**0.29**	**0.25**	0.95	2.30
c_3	2.50	2.01	1.95	1.69	1.20	2.40	**0.15**	**0.15**

Straight K-Means

0. *Data pre-processing.* Transform data into a quantitative matrix Y. This can be done according to the three step procedure described in section 2.4.
1. *Initial setting.* Choose the number of clusters, K, and tentative centroids $c_1, c_2, ..., c_K$, frequently referred to as seeds. Assume initial cluster lists S_k empty.
2. *Clusters update.* Given K centroids, determine clusters S'_k ($k = 1, ..., K$) with the Minimum distance rule.
3. *Stop-condition.* Check whether $S' = S$. If yes, end with clustering $S = S_k$, $c = (c_k)$. Otherwise, change S for S'.
4. *Centroids update.* Given clusters S_k, calculate within cluster means c_k ($k = 1, ..., K$) and go to Step 2.

This algorithm usually converges fast, depending on the initial setting. Location of the initial seeds may affect not only the speed of convergence but, more importantly, the final results as well. Let us give examples of how the initial setting may affect results.

Example 3.11. Successful application of K-Means

Let us apply K-Means to the same Masterpiece data in Table 3.2, this time starting with entities 2, 5 and 7 as tentative centroids (Step 1). To perform Step 2, the matrix of entity-to-centroid distances is computed (see Table 3.4 in which within column minima are boldfaced). The Minimum distance rule produces three cluster lists, $S_1 = \{1, 2, 3\}, S_2 = \{4, 5, 6\}$ and $S_3 = \{7, 8\}$. These coincide with the author-based clusters and produce within-cluster means (Step 4) already calculated in Table 3.1. Since these differ from the original tentative centroids (entities 2, 5, and 7), the algorithm returns to Step 2 of assigning clusters around the updated centroids. We do not do this here since the operation has been done already with distances in Table 3.3, which produced the same author-based lists according to the Minimum distance rule. The process thus stops. □

Example 3.12. Unsuccessful run of K-Means with different initial seeds

Let us take entities 1, 2 and 3 as the initial centroids (assuming the same data in Table 3.2). The Minimum distance rule, according to entity-to-centroid distances in

Table 3.4: Distances between entities 2, 5, 7 as seeds and the standardized Masterpiece entities.

Centroid	Row-point							
	1	2	3	4	5	6	7	8
2	**0.51**	**0.00**	**0.77**	1.55	1.82	2.99	1.90	2.41
5	2.20	1.82	1.16	**0.97**	**0.00**	**0.75**	0.83	1.87
7	2.30	1.90	1.81	1.22	0.83	1.68	**0.00**	**0.61**

Table 3.5: Distances between the standardized Masterpiece entities and entities 1, 2, 3 as seeds.

Centroid	Row-point							
	1	2	3	4	5	6	7	8
1	**0.00**	0.51	0.88	**1.15**	2.20	2.25	2.30	3.01
2	0.51	**0.00**	0.77	1.55	1.82	2.99	1.90	2.41
3	0.88	0.77	**0.00**	1.94	**1.16**	**1.84**	**1.81**	**2.38**

Table 3.5, leads to cluster lists $S_1 = \{1,4\}, S_2 = \{2\}$ and $S_3 = \{3,5,6,7,8\}$. With the centroids updated at Step 4 as means of these clusters, a new application of Step 3 leads to slightly changed cluster lists $S_1 = \{1,4,6\}, S_2 = \{2\}$ and $S_3 = \{3,5,7,8\}$. Their means calculated, it is not difficult to see that the Minimum distance rule does not change clusters anymore. Thus the lists represent the final outcome, which differs from the author-based solution.

□

The intuitive inappropriateness of the results in this example may be explained by the stupid choice of the initial centroids, all by the same author. However, K-Means can lead to inconvenient results even if the initial setting is selected according to clustering by authors.

Table 3.6: Distances between the standardized Masterpiece entities and entities 1, 4, 7 as tentative centroids.

Centroid	Row-point							
	1	2	3	4	5	6	7	8
1	**0.00**	**0.51**	**0.88**	1.15	2.20	2.25	2.30	3.01
4	1.15	1.55	1.94	**0.00**	0.97	**0.87**	1.22	2.46
7	2.30	1.90	1.81	1.22	**0.83**	1.68	**0.00**	**0.61**

Example 3.13. Unsuccessful K-Means with author-based initial seeds

With the initial centroids at rows 1, 4, and 7, the entity-to-centroid matrix in Table 3.6 leads to cluster lists $S_1 = \{1,2,3\}, S_2 = \{4,6\}$ and $S_3 = \{5,7,8\}$ that do not change in the follow-up operations. These results put a piece by Mark Twain among those by Leo Tolstoy. Not a good outcome. □

3.1.2 Square error criterion

The instability of clustering results with respect to the initial settings leads to a natural question whether there is anything objective in the method at all. Yes, there is.

It appears, there is a scoring function, an index, that is minimized by K-Means. To formulate the function, let us define the within cluster error. For a cluster S_k with centroid $c_k = (c_{kv})$, $v \in V$, its square error is defined as the summary distance from its elements to c_k:

$$W(S_k, c_k) = \sum_{i \in S_k} d(y_i, c_k) = \sum_{i \in S_k} \sum_{v \in V} (y_{iv} - c_{kv})^2. \tag{3.1}$$

The square error criterion is the sum of these values over all clusters:

$$W(S, c) = \sum_{k=1}^{K} W(S_k, c_k) \tag{3.2}$$

Criterion $W(S, c)$ (3.2) depends on two groups of arguments: cluster lists S_k and centroids c_k. Criteria of this type are frequently optimized with the so-called alternating minimization algorithm. This algorithm consists of a series of iterations. At each of the iterations, $W(S, c)$ is, first, minimized over S, given c, and, second, minimized over c, given the resulting S. This way, at each iteration a set c is transformed into a set c'. The calculations stop when c is stabilized, that is, $c' = c$.

Statement 3.3. *Straight K-Means is the alternating minimization algorithm for the summary square-error criterion (3.2) starting from seeds $c = \{c_1, ..., c_K\}$ specified in step 1.*

Proof: Equation

$$W(S, c) = \sum_{k=1}^{K} \sum_{i \in S_k} d(i, c_k),$$

following from (3.1), implies that, given $c = \{c_1, ..., c_K\}$, the Minimum distance rule minimizes $W(S, c)$ over S. Let us now turn to the problem of minimizing $W(S, c)$ over c, given S. It is obvious, that minimizing $W(S, c)$ over c can be done by minimizing $W(S_k, c_k)$ (3.1) over c_k independently for every $k = 1, ..., K$. Criterion $W(S_k, c_k)$ is a quadratic function of c_k and, thus, can be optimized with just first-order optimality conditions that the derivatives of $W(S_k, c_k)$ over c_{kv} must be equal to zero for all $v \in V$. These derivatives are equal to $F(c_{kv}) = -2\sum_{i \in S_k}(y_{iv} - c_{kv})$, $k = 1, ..., K; v \in V$. The condition $F(c_{kv}) = 0$ obviously leads to $c_{kv} = \sum_{i \in S_k} y_{iv}/|S_k|$, which proves that the optimal centroids must be within cluster gravity centers. This proves the statement.

Square-error criterion (3.2) is the sum of distances from entities to their cluster centroids. This can be reformulated as the sum of within cluster variances $\sigma_{kv}^2 = \sum_{i \in S_k}(y_{iv} - c_{kv})^2/N_k$ weighted by the cluster cardinalities:

$$W(S,c) = \sum_{k=1}^{K} \sum_{i \in S_k} \sum_{v \in V} (y_{iv} - c_{kv})^2 = \sum_{v \in V} \sum_{k=1}^{K} N_k \sigma_{kv}^2 \qquad (3.3)$$

Statement 3.3. implies, among other things, that K-Means converges in a finite number of steps because the set of all partitions S over a finite I is finite and $W(S,c)$ is decreased at each change of c or S. Moreover, as experiments show, K-Means typically does not move far away from the initial setting of c. Considered from the perspective of minimization of criterion (3.2), this leads to the conventional strategy of repeatedly applying the algorithm starting from various randomly generated sets of prototypes to reach as deep a minimum of (3.2) as possible. This strategy may fail especially if the feature set is large because in this case random settings cannot cover the space of solutions in a reasonable time.

Yet, there is a different perspective, of typology making, in which the criterion is considered not as something that must be minimized at any cost but rather a beacon for direction. In this perspective, the algorithm is a model for developing a typology represented by the prototypes. The prototypes should come from an external source such as the advice of experts, leaving to data analysis only their adjustment to real data. In such a situation, the property that the final prototypes are not far away from the original ones, is more of an advantage than not. What is important, though, is defining an appropriate, rather than random, initial setting.

The data recovery framework is consistent with this perspective since the model underlying K-Means is based on a somewhat simplistic claim that entities can be represented by their cluster's centroids, up to residuals. This model, according to section 5.2.1, leads to an equation involving K-Means criterion $W(S,c)$ (3.2) and the data scatter $T(Y)$:

$$T(Y) = B(S,c) + W(S,c) \qquad (3.4)$$

where

$$B(S,c) = \sum_{k=1}^{K} N_k c_{kv}^2 \qquad (3.5)$$

In this way, data scatter $T(Y)$ is decomposed into two parts: that one explained by the cluster structure (S,c), that is, $B(S,c)$, and the other unexplained, that is, $W(S,c)$. The larger the explained part the better the match between clustering (S,c) and data.

Criterion $B(S, c)$ measures the part of the data scatter taken into account by the cluster structure.

Example 3.14. Explained part of the data scatter
The explained part of the data scatter, $B(S, c)$, is equal to 43.7% of the data scatter $T(Y)$ for partition $\{\{1,4,6\},\{2\},\{3,5,7,8\}\}$, found with entities 1,2,3 as initial centroids. The score is 58.9% for partition $\{\{1,2,3\},\{4,6\},\{5,7,8\}\}$, found with entities 1,4,7 as initial centroids. The score is 64.0% for the author based partition $\{\{1,2,3\},\{4,5,6\},\{7,8\}\}$, which is thus superior. □

Advice for selecting the number of clusters and tentative centroids at Step 1 will be given in sections 3.2 and 7.5.

3.1.3 Incremental versions of K-Means

Incremental versions of K-Means are those at which Step 4, with its Minimum distance rule, is executed not for all of the entities but for one of them only. There can be two principal reasons for doing so:

R1 The user is not able to operate with the entire data set and takes entities in one by one, because of either the nature of the data generation process or the largeness of the data set sizes. The former cause is typical when clustering is done in real time as, for instance, in an on-line application. Under traditional assumptions of probabilistic sampling of the entities, convergence of the algorithm was explored in paper [83], from which K-Means became known publicly.

R2 The user operates with the entire data set, but wants to smooth the action of the algorithm so that no drastic changes in the cluster contents may occur. To do this, the user may specify an order of the entities and run entities one-by-one in this order for a number of times. (Each of the runs through the data set is referred to as an "epoch" in the neural network discipline.) The result of this may differ from that of Straight K-Means because of different computations. This computation can be especially effective if the order of entities is not constant but depends on their contributions to the criterion optimized by the algorithm. In particular, each entity $i \in I$ can be assigned value d_i, the minimum of distances from i to centroids $c_1, \ldots, c_K$, so that i minimizing d_i is considered first.

When an entity y_i joins cluster S_t whose cardinality is N_t, the centroid c_t changes to c'_t to follow the within cluster average values:

$$c'_t = \frac{N_t}{N_t+1}c_t + \frac{1}{N_t+1}y_i.$$

When y_i moves out of cluster S_t, the formula remains valid if all pluses are changed for minuses. By introducing the variable z_i which is equal to $+1$ when y_i joins the cluster and -1 when it moves out of it, the formula becomes

$$c'_t = \frac{N_t}{N_t + z_i} c_t + \frac{z_i}{N_t + z_i} y_i \tag{3.6}$$

Accordingly, the distances from other entities change to $d(y_j, c'_t)$.

Because of the incremental setting, the stopping rule of the straight version (reaching a stationary state) may be not necessarily applicable here. In case R1, the natural stopping rule is to end when there are no new entities observed. In case R2, the process of running through the entities one-by-one stops when all entities remain in their clusters. Also, the process may stop when a pre-specified number of runs (epochs) is reached.

This gives rise to the following version of K-Means.

Incremental K-Means: one entity at a time.
1. *Initial setting.* Choose the number of clusters, K, and tentative centroids, $c_1, c_2, ..., c_K$.
2. *Getting an entity.* Observe an entity $i \in I$ coming either randomly (setting R1) or according to a prespecified or dynamically changing order (setting R2).
3. *Cluster update.* Apply Minimum distance rule to determine to what cluster list S_t $(t = 1, ..., K)$ entity i should be assigned.
4. *Centroid update.* Update within cluster centroid c_t with formula (3.6). For the case in which y_i leaves cluster t' (in R2 option), $c_{t'}$ is also updated with (3.6). Nothing is changed if y_i remains in its cluster. Then the stopping condition is checked as described above, and the process moves to observing the next entity (Step 2) or ends (Step 5).
5. *Output.* Output lists S_t and centroids c_t with accompanying interpretation aids (as advised in section 3.4).

Example 3.15. Smoothing action of incremental K-Means

Let us apply version R2 to the Masterpieces data with the entity order dynamically updated and $K = 3$ starting with entities 1, 4 and 7 as centroids. Minimum distances d_i to the centroids for the five remaining entities are presented in the first column of Table 3.7 along with the corresponding centroid (iteration 0). Since $d_2 = 0.51$ is minimum among them, entity 2 is put in cluster I whose centroid is changed accordingly. The next column, iteration 1, presents minimum distances to the updated centroids.

This time the minimum is at $d_8 = 0.61$, so entity 8 is put in its nearest cluster III and its center is recomputed. In iteration 2, the distances are in column 2. Among remaining entities, 3, 5, and 6, the minimum distance is $d_3 = 0.70$, so 3 is added to its closest cluster I. Thus updated the centroid of cluster I leads to the change in minimum distances recorded at iteration 3. This time $d_6 = 0.087$ becomes minimum for the remaining entities 5 and 6 so that 6 joins cluster II and, in the next iteration,

Table 3.7: Minimum distances between standardized Masterpiece entities and dynamically changed centroids I, II and III.

Entity	Iteration					
	0	1	2	3	4	5
2	0.51/I	0.13/I	0.13/I	0.19/I	0.19/I	0.19/I
3	0.87/I	0.70/I	0.70/I	0.31 /I	0.31/I	0.31/I
5	0.83/III	0.83/III	0.97/II	0.97/II	0.97/II	0.28/II
6	0.87/II	0.87/II	0.87/II	0.87/II	0.22/II	0.25/II
8	0.61/III	0.61/III	0.15/III	0.15/III	0.15/III	0.15/III

5 follows it. Then the partition stabilizes: each entity is closer to its cluster centroid than to any other. The final partition of the set of masterpieces is the author based one. We can see that this procedure smoothes the process indeed: starting from the same centroids in Example 3.13, straight K-Means leads to a different and worse partition. □

3.2 Initialization of K-Means

To initialize K-Means, one needs to specify:

(1) the number of clusters, K, and

(2) initial centroids, $c_1, c_2, ..., c_K$.

Each of these can be of issue in practical computations. Both depend on the user's expectations related to the level of resolution and typological attitudes, which remain beyond the scope of the theory of K-Means. This is why some claim these considerations are beyond the clustering discipline. There have been however a number of approaches suggested for specifying the number and location of initial centroids, which will be briefly described in section 7.5.1. Here we present, first, the most popular existing approaches and, second, two approaches based on preliminary analysis of the data set structure.

3.2.1 Traditional approaches to initial setting

Conventionally, either of two extremes is adhered to in initial setting. One view assumes no knowledge of the data and domain and takes initial centroids randomly; the other, on the contrary, relies on the user being an expert and defining initial centroids as hypothetical prototypes.

The first approach randomly selects K of the entities (or generates K n-dimensional points within the feature ranges) as the initial seeds (centroids), and apply K-Means (either straight or incremental). After repeating this a pre-specified number of times (for instance, 100 or 1000), the best solution according to the square-error criterion (3.2) is taken. This approach can be handled by any package containing K-Means. For instance, SPSS allows the taking of the first K entities in a data set as the initial seeds. This can be repeated as many times as needed, each time reformatting the data matrix by putting a random K entity sample as its first K rows.

Selection of K can be done empirically by following this strategy for different values of K, say, in a range from 2 to 15. However, the optimal value of the square-error criterion decreases when K grows and thus cannot be utilized, as is, for the purpose. In the literature, a number of coefficients and tricks have been suggested based on the use of the square error (see later in section 7.5.1). Unfortunately, they all may fail even in the relatively simple situations of controlled computation experiments.

Comparing clusterings found for different K may lead to insights on the cluster structure. In many real world computations, the following phenomenon has been observed by the author and other researchers. When repeatedly proceeding from a larger K to $K-1$, the found $K-1$ clustering, typically, is rather similar to that found by merging some of clusters in the K clustering, in spite of the fact that the K- and $(K-1)$-clusterings are found independently. However, in the process of decreasing K this way, a critical value of K is reached such that $(K-1)$-clustering doesn't resemble K-clustering at all. If this is the case, the value of K can be taken as that corresponding to the cluster structure.

This can be a viable strategy. There are two critical points though.

1. The K-Means algorithm, as is, doesn't seek a global minimum of the square-error criterion and, moreover, the local minima achieved with K-Means are not very deep. Thus, with the number of entities in order of hundreds or thousands and K within a dozen, or more, the number of tries needed to reach a representative set of the initial centroids may become too large and make it a computationally challenging problem.

   To overcome this, some effective computational strategies have been suggested as that of random jumps from a subset of centroids. Such random track changing, typically, produces much deeper minima than the standard K-Means [45].

2. Even if one succeeds in getting a deep or a global minimum of the square-error criterion, it should not be taken for granted that the clusters found reflect the cluster structure. There are some intrinsic flaws in the criterion that would not allow us to accept it as the only means for deciding upon whether the clusters minimizing it are those we are looking for. The square-error criterion needs to be supplemented with other tools for getting better insights into the data structure. Setting of the initial centroids can be utilized as such a tool.

The other approach relies on the opinion of an expert in the subject domain.

Example 3.16. K-Means at Iris data

Table 3.8 presents results of the straight K-Means applied to the Iris data on page 11 with K=3 and specimens numbered 1, 51, and 101 taken as the initial centroids and

Table 3.8: Cross-classification of 150 Iris specimens according to K-Means clustering and the genera; entries show Count/Proportion.

Cluster	Iris genus Setosa	Versicolor	Virginica	Total
S_1	50/0.333	0/0	0/0	50/0.333
S_2	0/0	47/0.313	14/0.093	61/0.407
S_3	0/0	3/0.020	36/0.240	39/0.260
Total	50/0.333	50/0.333	50/0.333	150/1.000

cross-classified with the prior three class partition. The clustering does separate genus *Setosa* but misplaces 14+3=17 specimens between two other genera. This corresponds to the visual pattern in Figure 1.10, page 25. □

Similarly, an expert may propose to distinguish numeral digits by the presence of a closed drawing in them, so that this feature is present in 6 and absent from 1, and suggest these entities as the initial seeds. The expert may even go further and suggest one more feature, presence of a semi-closed drawing instantiated by 3, to be taken into account.

This is a viable approach, too. It allows seeing how the conceptual types relate to the data and to what extent the hypothetical seed combinations match real data.

However, in a common situation in which the user cannot make much sense of his data because they reflect superficial measurable features rather than those of essence, which cannot be measured, the expert vision may fail to suggest a reasonable degree of resolution, and the user should take a more data-driven approach to tackle the problem. Two data-driven approaches are described in the next two sections.

3.2.2 MaxMin for producing deviate centroids

This approach is based on the following intuition. If there are cohesive clusters in the data, then entities within any cluster must be close to each other and rather far away from entities in other clusters. The following method, based on this intuition, has proved to work well in real and simulated experiments.

MaxMin
1. Take entities $y_{i'}$ and $y_{i''}$ maximizing the distance $d(y_i, y_j)$ over all $i, j \in I$ as c_1 and c_2.
2. For each of the entities y_i, that have not been selected to the set c of initial seeds so far, calculate $d_c(y_i)$, the minimum of its distances to $c_t \in c$.
3. Find i^* maximizing $d_c(y_i)$ and check Stop-condition (see below). If it doesn't hold, add y_{i^*} to c and go to Step 2. Otherwise, end and output c as the set of initial seeds.

As the Stop-condition in MaxMin either or all of the following pre-specified constraints can be utilized:

1. The number of seeds has reached a pre-specified threshold.
2. Distance $d_c(y_{i^*})$ is larger than a pre-specified threshold such as $d = d(c_1, c_2)/3$.
3. There is a significant drop, such as 35%, in the value of $d_c(y_{i^*})$ in comparison to that at the previous iteration.

Example 3.17. MaxMin for selecting intial seeds
The table of entity-to-entity distances for Masterpieces is displayed in Table 3.9. The maximum distance here is 3.43, between AK and YA, which makes the two of

Table 3.9: Distances between Masterpieces from Table 3.2.

	OT	DS	GE	TS	HF	YA	WP	AK
OT	0.00	0.51	0.88	1.15	2.20	2.25	2.30	3.01
DS	0.51	0.00	0.77	1.55	1.82	2.99	1.90	2.41
GE	0.88	0.77	0.00	1.94	1.16	1.84	1.81	2.38
TS	1.15	1.55	1.94	0.00	0.97	0.87	1.22	2.46
HF	2.20	1.82	1.16	0.97	0.00	0.75	0.83	1.87
YA	2.25	2.99	1.84	0.87	0.75	0.00	1.68	3.43
WP	2.30	1.90	1.81	1.22	0.83	1.68	0.00	0.61
AK	3.01	2.41	2.38	2.46	1.87	3.43	0.61	0.00

them initial centroids according to MaxMin. The distances from other entities to these two are in Table 3.10; those minimal at the two are boldfaced. The maximum among them, the next MaxMin distance, is 2.41 between DS and AK. The decrease here is less than 30% suggesting that this can represent a different cluster. Thus, we

Table 3.10: Distances from Masterpieces entities to YA and AK

	OT	DS	GE	TS	HF	WP
YA	**2.25**	3.00	**1.84**	**0.87**	**0.75**	1.68
AK	3.01	**2.41**	2.38	2.46	1.87	**0.61**

add DS to the list of candidate centroids and then need to look at distances from other entities to these three (see Table 3.11). This time the MaxMin distance is 0.87 between TS and YA. We might wish to stop the process at this stage since we expect only three meaningful clusters in Masterpieces data and, also, there is a significant drop, 64% of the previous MaxMin distance. It is useful to remember that such a clear-cut situation may not necessarily occur in other examples. The three seeds selected have been shown in previous examples to produce the author based clusters with K-Means. □

Table 3.11: Distances between DS, YA, and AK and other Masterpiece entities.

	OT	GE	TS	HF	WP
DS	**0.51**	**0.77**	1.55	1.82	1.90
YA	2.25	1.84	**0.87**	**0.75**	1.68
AK	3.01	2.38	2.46	1.87	**0.61**

The issues related to this approach are typical in data mining. First, it involves ad hoc thresholds which are not substantiated in terms of data. Second, it can be computationally intensive when the number of entities N is large since finding the maximum distance at Step 1 involves computation of $O(N^2)$ distances.

3.2.3 Deviate centroids with Anomalous pattern

The method described in this section provides an alternative to MaxMin for the initial setting, which is less intensive computationally and, also, reduces the number of ad hoc parameters.

Reference point based clustering

To avoid the computationally intensive problems of analyzing pair-wise distances, one may employ the concept of a reference point which is chosen to exemplify an average or norm of the features which define the entities. For example, the user might choose, as representing a "normal student," a point which indicates good marks in tests and serious work in projects, and then see what patterns of observed behavior deviate from this. Or, a bank manager may set as his reference point, a customer having specific assets and backgrounds, not necessarily averaged, to see what types of customers deviate from this. In engineering, a moving robotic device should be able to classify the elements of the environment according to the robot's location, with things that are closer having more resolution, and things that are farther having less resolution: the location is the reference point in this case. In many cases the gravity center of the entire entity set can be the reference point of choice.

Availability of a reference point allows the comparison of entities with it, not with each other, which drastically reduces computations. To find a cluster which is most distant from a reference point, a version of K-Means described in [92] can be utilized. According to this procedure, the only ad hoc choice is the cluster's seed. There are two seeds here: the reference point which is unvaried in the process and the cluster's seed, which is taken to be the entity which is farthest from the reference point. Only the anomalous cluster is built here, defined as the set of points that are closer to the cluster seed than to the reference point. Then the cluster seed is substituted by the cluster's gravity center, and the procedure is reiterated until it converges. An exact formulation of this follows.

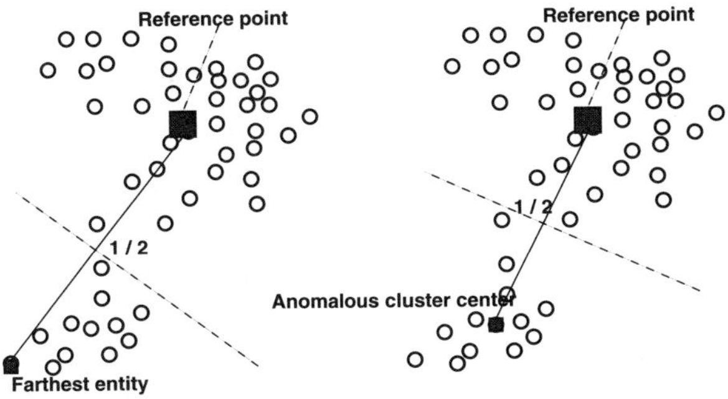

Figure 3.1: Extracting an 'Anomalous pattern' cluster with the reference point in the gravity center: the initial iteration is on the left and the final one on the right.

Anomalous pattern (AP)
1. *Pre-processing.* Specify a reference point $a = (a_1, ..., a_n)$ (this can be the data grand mean) and standardize the original data table with formula (2.22) at which shift parameters a_k are the reference point coordinates. (This way, the space origin is shifted into a.)
2. *Initial setting.* Put a tentative centroid, c, as an entity which is the most distant from the origin, 0.
3. *Cluster update.* Determine cluster list S around c against the only other "centroid" 0 with the Minimum distance rule so that y_i is assigned to S if $d(y_i, c) < d(y_i, 0)$.
4. *Centroid update.* Calculate the within S mean c' and check whether it differs from the previous centroid c. If c' and c do differ, update the centroid by assigning $c \leftarrow c'$ and return to Step 3. Otherwise, go to 5.
5. *Output.* Output list S and centroid c with accompanying interpretation aids (as advised in section 3.4) as the most anomalous pattern.

The process is illustrated in Figure 3.1. Obviously, the Anomalous pattern method is a version of K-Means in which:

(i) the number of clusters K is 2;

(ii) centroid of one of the clusters is 0, which is forcibly kept there through all the iterations;

(iii) the initial centroid of the anomalous cluster is taken as an entity point which is the most distant from 0.

Property (iii) mitigates the issue of determining appropriate initial seeds, which allows using Anomalous pattern algorithm for finding an initial setting for K-Means.

Like K-Means itself, the Anomalous pattern alternately minimizes a criterion,

$$W(S,c) = \sum_{i \in S} d(y_i, c) + \sum_{i \notin S} d(y_i, 0) \tag{3.7}$$

which is a specific version of K-Means general criterion $W(S,c)$ in (3.2): S is a partition in the general criterion and a subset in AP. More technical detail of the method can be found in section 5.5.

Example 3.18. Anomalous pattern in Market towns
The Anomalous pattern method can be applied to the Market towns data in Table 1.1 assuming the grand mean as the reference point and scaling by range. The point farthest from 0, the tentative centroid at step 2, appears to be entity 35 (St Austell) whose distance from zero is 4.33, the maximum. Step 3 adds three more entities, 26, 29 and 44 (Newton Abbot, Penzance and Truro), to the cluster. They are among the largest towns in the data, though there are some large towns like Falmouth that are out of the list, thus being closer to 0 rather than to St Austell in the range standardized feature space. After one more iteration, the anomalous cluster stabilizes.

Table 3.12: Iterations in finding an anomalous pattern in Market towns data.

Iteration	List	#	Distance	Cntr	Cntr, %
1	26, 29, 35, 44	4	2.98	11.92	28.3
2	4, 9, 25, 26, 29, 35, 41, 44	8	1.85	14.77	35.1

The iterations are presented in Table 3.12. It should be noted that the scatter's cluster part (contribution) increases along the iterations as follows from the theory in section 5.5.3: the decrease of the distance between centroid and zero is well compensated by the influx of entities. The final cluster consists of 8 entities and takes into account 35.13 % of the data scatter. Its centroid is displayed in Table 3.13. As frequently happens, the anomalous cluster here consists of better off entities – towns with all the standardized centroid values larger than the grand mean by 30 to 50 per cent of the feature ranges. This probably relates to the fact that they comprise eight out of eleven towns which have a resident population greater than 10,000. The other three largest towns have not made it into the cluster because of their deficiencies in services such as Hospitals and Farmers' markets. The fact that the scale of measurement of population is by far the largest in the original table doesn't much affect the computation here as it runs with the range standardized scales at which the total contribution of this feature is mediocre, about 8.5% only (see Table 2.18). It is rather a concerted action of all features associated with greater population which makes up the cluster. As follows from the last line in Table 3.13, the most important for the cluster separation

Table 3.13: Centroid of the extracted pattern of Market towns.

Centroid	P	PS	Do	Ho	Ba	SM	Pe	DIY	SP	PO	CAB	FM
Real	18484	7.6	3.6	1.1	11.6	4.6	4.1	1.0	1.4	6.4	1.2	.4
Stded	.51	.38	.56	.36	.38	.38	.30	.26	.44	.47	.30	.18
Over GM, %	151	152	163	181	170	139	102	350	181	143	94	88
Related Cntr, %	**167**	147	**154**	81	144	126	84	87	104	**163**	51	10

are the following features: Population, Post offices, and Doctors, highlighted with the boldface. This analysis suggests a simple decision rule separating the cluster entities from the rest with these variables: "P is greater than 10,000 and Do is 3 or greater." □

3.3 Intelligent K-Means

3.3.1 Iterated Anomalous pattern for iK-Means

When clusters in the feature space are well separated from each other or the cluster structure can be thought of as a set of differently contributing clusters, the clusters can be found with iterative application of Anomalous pattern that mitigates the need for pre-setting the number of clusters and their initial centroids. Moreover, this can be used as a procedure to meaningfully determine the number of clusters and initial seeds for K-Means. In this way we come to an algorithm that can be referred to as an intelligent K-Means, because it relieves from the user the task of specifying the initial setting.

Some other potentially useful features of the method relate to its flexibility with regard to dealing with outliers and the "swamp" of inexpressive, ordinary, entities around the grand mean.

iK-Means

0. *Setting.* Put $t = 1$ and I_t the original entity set. Specify a threshold of resolution to discard all AP clusters whose cardinalities are less than the threshold.
1. *Anomalous pattern.* Apply AP to I_t to find S_t and c_t. There can be either option taken: do Step 1 of AP (standardization of the data) at each t or only at $t = 1$. The latter is the recommended option as it is compatible with the theory in section 5.5.
2. *Control.* If Stop-condition (see below) does not hold, put $I_t \leftarrow I_t - S_t$ and $t \leftarrow t + 1$ and go to Step 1.
3. *Removal of small clusters.* Remove all of the found clusters that are smaller than a pre-specified *cluster discarding threshold* for the cluster size. (Entities comprising singleton clusters should be checked for the errors in their data entries.) Denote the number of remaining clusters by K and their centroids by c_1,..., c_K.
4. *K-Means.* Do Straight (or Incremental) K-Means with c_1,..., c_K as initial seeds.

The Stop-condition in this method can be any or all of the following:

1. **All clustered.** $S_t = I_t$ so that there are no unclustered entities left.
2. **Large cumulative contribution.** The total contribution of the first t

clusters to the data scatter has reached a pre-specified threshold such as 50 %.

3. **Small cluster contribution.** Contribution of t-th cluster is too small, say, compared to the order of average contribution of a single entity, $1/N$.

4. **Number of clusters reached.** Number of clusters, t, has reached a pre-specified value K.

The first condition is natural if there are "natural" clusters that indeed differ in their contributions to the data scatter. The second and third conditions can be considered as imposing further degrees of resolution with which the user looks at the data.

At step 4, K-Means can be applied to either the entire dataset or to the set from which the smaller clusters have been removed. This may depend on the situation: in some problems, such as structuring of a set of settlements for better planning or monitoring, no entity should be left out of the consideration, whereas in other problems, such as developing synoptic descriptions for text corpora, some deviant texts should be left out of the coverage.

Example 3.19. Iterated Anomalous patterns in Market towns

Applied to the Market towns data with Stop-condition 1, the iterated AP algorithm has produced 12 clusters of which 5 are singletons. Each of the singletons has a strange pattern of town facilities with no similarity to any other town in the list. For instance, entity 19 (Liskeard, 7044 residents) has an unusually large number of Hospitals (6) and CABs(2), which makes it a singleton cluster.

The seven non-singleton clusters are in Table 3.14, in the order of their extraction in the iterated AP. Centroids of the seven clusters are presented in Table 3.20 in the next section.

Table 3.14: Iterated AP Market towns clusters.

Cluster	Size	Elements	Cntr,%
1	8	4, 9, 25, 26, 29, 35, 41, 44	35.1
3	6	5, 8 , 12, 16, 21, 43	10.0
4	18	2, 6, 7, 10, 13, 14, 17, 22, 23, 24, 27, 30, 31, 33, 34, 37, 38, 40	18.6
5	2	3 , 32	2.4
6	2	1,11	1.6
8	2	39 , 42	1.7
11	2	20 45	1.2

The cluster structure doesn't much change when, according to the iK-Means algorithm, Straight K-Means is applied to the seven centroids (with the five singletons put

Table 3.15: Clusters found by the iterated AP algorithm in Bribery data.

Cluster	Size	Elements	Contribution, %
1	7	5,16,23,27,28,41,42	9.8
2	1	25	2.2
3	2	17,22	3.3
4	1	49	2.2
5	1	2	2.1
6	1	35	2.1
7	13	12,13,20,33,34,38,39,43 45,47,48,50,51	10.7
8	9	4,6,9,10,21,26,30,31,40	10.2
9	5	1,3,15,29,32	6.3
10	2	7,52	3.3
11	3	8,14,36	3.4
12	8	11,24,37,44,46,53,54,55	7.8
13	2	18,19	2.6

back into the data). Moreover, similar results have been observed with clustering of the original list of about thirteen hundred Market towns described by an expanded list of eighteen characteristics of their development: the number of non-singleton clusters was the same, with their descriptions (see page 101) very similar.

□

Example 3.20. Intelligent K-Means on Bribery data

Let us apply iK-Means to the Bribery data in Table 1.12 on page 20. According to the prescriptions above, the data processing includes the following steps:

1. Data standardization. This is done by subtracting the feature averages (grand means) from all entries and then dividing them by the feature ranges. For a binary feature corresponding to a qualitative category, this reduces to subtraction of the category proportion, p, from all the entries which in this way become either $1-p$, for "yes," and $-p$, for "no."

2. Repeatedly performing AP clustering. Applying AP to the pre-processed data matrix with the reference point taken as the space origin 0 and never altered, 13 clusters have been produced as shown in Table 3.15. They explain 64% of the data variance.

3. Initial setting for K-Means. There are only 5 clusters that have more than three elements according to Table 3.15. This defines the number of clusters as well as the initial setting: the first elements of the five larger clusters, indexed as 5, 12, 4, 1, and 11, are taken as the initial centroids.

4. Performing K-Means. K-Means, with the five centroids from the previous step, produces five clusters presented in Table 3.16. They explain 45% of the data scatter. The reduction of the proportion of the explained data scatter is obviously caused by the reduced number of clusters.

Conceptual description of the clusters is left to the next section (see page 106) which is devoted to interpretation aids. □

Table 3.16: Clusters found by K-Means in the entire Bribery data set from the largest clusters in Table 3.15.

Cluster	#	Elements	Contribution, %
1	8	5,16,23,25,27,28,41,42	10.0
2	19	7,8,12,13,14,20,33,34,35,3638,39,43,45,47,48,50,51,52	9.8
3	10	4,6,9,10,21,22,26,30,31,40	10.0
4	7	1,3,15,17,29,32,49	7.0
5	11	2,11,18,19,24,37,44,46,53,54,55	8.1

3.3.2 Cross validation of iK-Means results

As described in section 1.2.6, the issue of validation of clusters may be subject to different perspectives. According to the classification paradigm, validation of clusters is provided by their interpretation, that is, by the convenience of the clusters and their fitting into and enhancing the existing knowledge. In the statistics paradigm, a cluster structure is validated by its correspondence to the underlying model. In the machine learning perspective, it is learning algorithms that are to be validated. In data mining, one validates the cluster structure found. In machine learning and data mining, validation is treated as the testing of how stable the algorithm results are with respect to random changes in the data. We refer the reader to section 7.5 for a general discussion of validation criteria in clustering.

Here we concentrate on the most popular validation method, m-fold cross-validation. According to this method, the entity set is randomly partitioned into m equal parts and m pairs of training and testing sets are formed by taking each one of the m parts as the testing set, with the rest considered the training set.

This scheme is easy to use regarding the problems of learning of decision rules: a decision rule is formed using a training set and then tested on the corresponding testing set. Then testing results are averaged over all m train-test experiments. How can this line of thought be applied to clustering?

In the literature, several methods for extending of the cross-validation techniques to clustering have been described (see references in section 7.5.2). Some of them fall in the machine learning perspective and some in the data mining perspective. The common idea is that the set of m training sets supplied by the cross validation approach constitute a convenient set of random samples from the entity set. In the remainder of this section, we describe somewhat simplified experiments in each of the two frameworks.

In the machine learning framework, one tests the consistency of a clustering algorithm. To do this, results of the algorithm run over each of the m training sets are compared. But how can two clusterings be compared if they partition different sets? One way to do this is by extending each clustering from the training sct to thc full cntity sct by assigning appropriate cluster labels to the test set elements. Another way would be to compare partitions pairwise over the overlap of their training sets. The overlap is not necessarily small. If, for instance, $m = 10$, then each of the training sets covers 90% of entities and the pairwise overlap is 80%.

In data mining, it is the clustering results that are tested. In this framework, the selected clustering method is applied to the entire data set before the set is split into m equal-size parts. Then m training sets are formed as usual, by removing one of the parts and combining the other parts. These training sets are used to verify the clustering results found on the entire data set. To do this, the clustering algorithm is applied to each of the m training sets and the found clustering is compared with that obtained on the entire data set.

Let us consider, with examples, how these strategies can be implemented.

Example 3.21. Cross-validation of iK-Means clusters of the Market towns data

Let us address the issue of consistency of clustering results, a data mining approach. We already have found a set of clusters in the Market towns data, see example 3.19 on page 94. This will be referred to as base clustering. To explore how stable base clusters are, let us do 10-fold cross-validation. First, randomly partition the set of 45 towns in 10 classes of approximately the same size, five classes with four towns and five classes with five towns in each. Taking out each of the classes, we get ten 90% subsamples of the original data as the training sets and run iK-Means on each of them. To see how much these clusterings differ from the base clustering found using the entire set, we use three scoring functions, as follows.

1. **Average distance between centroids** *adc*. Let c_k $(k = 1, ..., 7)$ be base centroids and c'_l $(l = 1, ..., L)$ centroids of the clustering found on a 90% sample. For each c_k find the nearest c'_l over $l = 1, ..., L$, calculate $d(c_k, c'_l)$ and average the distance over all $k = 1, ..., 7$. (The correspondence between c_k and c'_l can also be established with the so-called best matching techniques [3].) This average distance scores the difference between base clusters and sample clusters. The smaller it is the more consistent is the base clustering.
2. **Relative distance between partitions of samples** M. Given a 90% training sample, let us compare two partitions of it: (a) the partition found on it with the clustering algorithm and (b) the base partition constrained to the sample. Cross classifying these two partitions, we get a contingency table $P = (p_{tu})$ of frequencies p_{tu} of sample entities belonging to the t-th class of one partition and the u-th class of the other. The distance, or mismatch coefficient, is

$$M = \sum_t p_{t+}^2 + \sum_u p_{+u}^2 - 2\sum_{t,u} p_{tu}^2$$

where p_{t+} and p_{+u} are summary frequencies over rows and columns of P, as introduced later in formula (7.12).
3. **Relative chi-square contingency coefficient** T. This is computed in the same way as distance M; the only difference is that now chi-squared coefficient (2.12), (2.13)

$$X^2 = \sum_{t,u} p_{tu}^2/(p_{t+}p_{+u}) - 1$$

Table 3.17: Averaged results of fifteen cross-validations of Market towns clusters with real and random data.

Method	Real data	Random data
adc	0.064 (0.038)	0.180 (0.061)
M_s	0.018 (0.018)	0.091 (0.036)
T	0.865 (0.084)	0.658 (0.096)

and its normalized version $T = X^2/\sqrt{(K-1)(L-1)}$, the Tchouprov coefficient, are used. Tchouprov coefficient cannot be greater than 1.

Averaged results of fifteen independent 10-fold cross validation tests are presented in the left column of Table 3.17; the standard deviations of the values are in parentheses.

We can see that distances *adc* and M_s are low and contingency coefficient T is high. But how low and how high are they? Can any cornerstones or benchmarks be found?

One may wish to compare *adc* with the average distance between uniformly random vectors. This is not difficult, because the average squared difference $(x-y)^2$ between numbers x and y that are uniformly random in a unity interval is 1/6. This implies that the average distance in 12-dimensional space is 2 which is by far greater than the observed 0.064.

This difference however, shouldn't impress anybody, because the distance 2 refers to an unclustered set. Let us generate thus a uniformly random 45×12 data table and simulate the same computations as with the real data. Results of these computations are in the column on the right of Table 3.17. We can see that distances *adc* and M_s over random data are small too; however, they are 3-5 times greater than those on the real data. If one believes that the average distances at random and real data may be considered as sampling averages of normal or chi-square distributions, one may consider a statistical test of difference such as that by Fisher [63, 50] to be appropriate and lead to a statistically sound conclusion that the hypothesis that the clustering of real data differs from that of random data can be accepted with a great confidence level. □

Example 3.22.

Cross-validation of iK-Means algorithm on the Market towns data

In this example, the cross-validation techniques are applied within the machine learning context, that is to say, we are going to address the issue of the consistency of the clustering algorithm rather than its results.

Thus, the partitions found on the training samples will be compared not with the base clustering but with each other. A 10-fold cross-validation is applied here as in the previous example. Ten 90% cross-validation subsamples of the original data are produced and iK-Means is applied to each of them. Two types of comparison between the ten subsample partitions are used, as follows.

1. **Comparing partitions on common parts.** Two 90% training samples' overlap comprises 80% of the original entities, which allows the building of their contingency table over those common entities. Then both the distance M and chi-squared T coefficients can be used.
2. **Comparing partitions by extending them to the entire entity set.** Given a 90% training sample, let us first extend it to the entire entity set. To

Table 3.18: Averaged comparison scores between iK-Means results at 80% real Market towns and random data.

Method	Real data	Random data
M_s	0.027 (0.025)	0.111 (0.052)
T	0.848 (0.098)	0.604 (0.172)

Table 3.19: Averaged comparison scores between iK-Means results extended to all real Market towns and random data.

Method	Real data	Random data
M_s	0.032 (0.028)	0.128 (0.053)
T	0.832 (0.098)	0.544 (0.179)

do so, each entity from the 10% testing set is assigned to the cluster whose centroid is the nearest to the entity. Having all ten 90% partitions extended this way to the entire data set, their pair-wise contingency tables are built and scoring functions, the distance M and chi-squared T coefficients, are calculated.

Tables 3.18 and 3.19 present results of the pair-wise comparison between partitions found by iK-Means applied to the Market towns data in both ways, on 80% overlaps and on the entire data set after extension, averaged over fifteen ten-fold cross-validation experiments. The cluster discarding threshold has been set to 1 as in the previous examples. We can see that these are similar to figures observed in the previous example though the overall consistency of clustering results decreases here, especially when comparisons are conducted over extended partitions.

It should be noted that the issue of consistency of the algorithm is treated somewhat simplistically in this example, with respect to the Market towns data only, not to a pool of data structures. Also, the concept of algorithm's consistency can be defined differently, for instance, with regards to the criterion optimized by the algorithm.

□

Example 3.23. Higher dimensionality effects

It is interesting to mention that applying the same procedure to the original set of 18 features (not presented), the following phenomenon has been observed. When a matrix 45×18 is filled in by a set of uniformly random numbers, iK-Means with the cluster discarding threshold 2, produces two clusters only. However, at the 90% training subsamples iK-Means fails most of the times to produce more than one non-trivial cluster. This is an effect of the higher dimensionality of the feature space relative to the number of entities in this example. Random points are situated too far away from each other in this case and can not be conflated by iK-Means into clusters. One may safely claim that iK-Means differs from other clustering algorithms in that respect that, in contrast to the others, it may fail to partition a data set if it is random. This happens not always but only in the cases in which the number of features is comparable to or greater than half of the number of entities. □

3.4 Interpretation aids

As it was already pointed out, interpretation is an important part of clustering, especially from the classification perspective in which it is a validation tool as well. Unfortunately, this subject is generally not treated within the same framework as 'proper' clustering. The data recovery view of clustering allows us to fill in some gaps here as described in this section.

3.4.1 Conventional interpretation aids

Two conventional tools for interpreting K-Means clustering results (S, c) are:

(1) analysis of cluster centroids c_t and

(2) analysis of bivariate distributions between cluster partition $S = \{S_t\}$ and various features.

In fact, under the zero-one coding system for categories, cross-classification frequencies are nothing but cluster centroids, which allows us to safely suggest that analysis of cluster centroids at various feature spaces is the only conventional interpretation aid.

Example 3.24. Conventional interpretation aids applied to Market towns clusters.

Let us consider Table 3.20 displaying centroids of the seven clusters of Market towns data both in real and range standardized scales. These show some tendencies rather clearly. For instance, the first cluster appears to be a set of larger towns that score 30 to 50 % higher than average on almost all 12 features in the feature space. Similarly, cluster 3 obviously relates to smaller than average towns. However, in other cases, it is not always clear what features caused the separation of some clusters. For instance, both clusters 6 and 7 seem too close to the average to have any real differences at all. □

Table 3.20: Patterns of Market towns in the cluster structure found with iK-Means; the first column displays cluster numbering (top) and cardinalities (bottom).

k/#	Centr	P	PS	Do	Ho	Ba	Su	Pe	DIY	SP	PO	CAB	FM
1	Real	18484	7.63	3.63	1.13	11.63	4.63	4.13	1.00	1.38	6.38	1.25	0.38
8	Stand	0.51	0.38	0.56	0.36	0.38	0.38	0.30	0.26	0.44	0.47	0.30	0.17
2	Real	5268	2.17	0.83	0.50	4.67	1.83	1.67	0.00	0.50	1.67	0.67	1.00
6	Stand	-0.10	-0.07	-0.14	0.05	0.02	-0.01	-0.05	-0.07	0.01	-0.12	0.01	0.80
3	Real	2597	1.17	0.50	0.00	1.22	0.61	0.89	0.00	0.06	1.44	0.11	0.00
18	Stand	-0.22	-0.15	-0.22	-0.20	-0.16	-0.19	-0.17	-0.07	-0.22	-0.15	-0.27	-0.20
4	Real	11245	3.67	2.00	1.33	5.33	2.33	3.67	0.67	1.00	2.33	1.33	0.00
3	Stand	0.18	0.05	0.16	0.47	0.05	0.06	0.23	0.15	0.26	-0.04	0.34	-0.20
5	Real	5347	2.50	0.00	1.00	2.00	1.50	2.00	0.00	0.50	1.50	1.00	0.00
2	Stand	-0.09	-0.04	-0.34	0.30	-0.12	-0.06	-0.01	-0.07	0.01	-0.14	0.18	-0.20
6	Real	8675	3.80	2.00	0.00	3.20	2.00	2.40	0.00	0.00	2.80	0.80	0.00
5	Stand	0.06	0.06	0.16	-0.20	-0.06	0.01	0.05	-0.07	-0.24	0.02	0.08	-0.20
7	Real	5593	2.00	1.00	0.00	5.00	2.67	2.00	0.00	1.00	2.33	1.00	0.00
3	Stand	-0.08	-0.09	-0.09	-0.20	0.04	0.10	-0.01	-0.07	0.26	-0.04	0.18	-0.20

3.4.2 Contribution and relative contribution tables

Here two more interpretation aids are proposed:

1. Decomposition of the data scatter over clusters and features (table ScaD);

2. Quetelet coefficients for the decomposition (table QScaD).

According to (3.4) and (3.5), clustering decomposes the data scatter $T(Y)$ in the explained and unexplained parts, $B(S,c)$ and $W(S,c)$, respectively. The explained part can be further presented as the sum of additive items $B_{kv} = N_k c_{kv}^2$, which account for the contribution of every pair S_k $(k = 1, ..., K)$ and $v \in V$, a cluster and a feature. The unexplained part can be further additively decomposed in contributions $W_v = \sum_{k=1}^{K} \sum_{i \in S_k} (y_{iv} - c_{kv})^2$, which can be differently expressed as $W_v = T_v - B_{+v}$ where T_v and B_{+v} are parts of $T(Y)$ and $B(S,c)$ related to feature $v \in V$, $T_v = \sum_{i \in I} y_{iv}^2$ and $B_{+v} = \sum_{k=1}^{K} B_{kv}$.

This can be displayed as a decomposition of $T(Y)$ in a table ScaD whose rows correspond to clusters, columns to variables and entries to the contributions (see Table 3.21).

Table 3.21: ScaD: Decomposition of the data scatter over a K-Means cluster structure.

Feature Cluster	$f1$	$f2$	fM	Total
S_1	B_{11}	B_{12}	B_{1M}	B_{1+}
S_2	B_{21}	B_{22}	B_{2M}	B_{2+}
S_K	B_{K1}	B_{K2}	B_{KM}	B_{K+}
Expl	B_{+1}	B_{+2}	B_{+M}	$B(S,c)$
Unex	W_1	W_2	W_M	$W(S,c)$
Total	T_1	T_2	T_M	$T(Y)$

Summary rows, Expl and Total, and column, Total, are added to the table; they can be expressed as percentages of the data scatter $T(Y)$. The notation follows the notation of flow data. The row Unex accounts for the "unexplained" differences $W_v = T_v - B_{+v}$. The contributions highlight relative roles of features both at individual clusters and in total.

These can be applied within clusters as well (see Table 3.26 further on as an example).

Example 3.25. Contribution table ScaD for Market towns clusters

Table 3.22 presents the Market towns data scatter decomposed, as in Table 3.21, over both clusters and features.

The table shows that, among the variables, the maximum contribution to the data scatter is reached at FM. This can be attributed to the fact that FM is a binary

Table 3.22: Table ScaD at Market towns: Decomposition of the data scatter over clusters and features.

Cl-r	P	PS	Do	Ho	Ba	Su	Pe	DIY	SP	PO	CAB	FM	Total	Tot.,%
1	2.09	1.18	2.53	1.05	1.19	1.18	0.71	0.54	1.57	1.76	0.73	0.24	14.77	35.13
2	0.06	0.03	0.11	0.01	0.00	0.00	0.02	0.03	0.00	0.09	0.00	3.84	4.19	9.97
3	0.86	0.43	0.87	0.72	0.48	0.64	0.49	0.10	0.85	0.39	1.28	0.72	7.82	18.60
4	0.10	0.01	0.07	0.65	0.01	0.01	0.16	0.07	0.20	0.00	0.36	0.12	1.75	4.17
5	0.02	0.00	0.24	0.18	0.03	0.01	0.00	0.01	0.00	0.04	0.06	0.08	0.67	1.59
6	0.02	0.02	0.12	0.20	0.02	0.00	0.01	0.03	0.30	0.00	0.03	0.20	0.95	2.26
7	0.02	0.02	0.03	0.12	0.00	0.03	0.00	0.02	0.20	0.00	0.09	0.12	0.66	1.56
Expl	3.16	1.69	3.96	2.94	1.72	1.88	1.39	0.79	3.11	2.29	2.56	5.33	30.81	73.28
Unex	0.40	0.59	0.70	0.76	0.62	0.79	1.02	0.96	1.20	0.79	1.52	1.88	11.23	26.72
Total	3.56	2.28	4.66	3.70	2.34	2.67	2.41	1.75	4.31	3.07	4.08	7.20	42.04	100.00

variable: as shown in section 2.1.2, contributions of binary variables are maximal when they cover about half of the sample. The least contributing is DIY. The value of the ratio of the explained part of DIY to the total contribution, 0.79/1.75=0.451, amounts to the correlation ratio between the partition and DIY, as explained in sections 3.4.4 and 5.2.3.

The entries in the table actually combine together cardinalities of clusters with squared differences between the grand mean vector and within-cluster centroids. Some show an exceptional value such as contribution 3.84 of FM to cluster 2, which covers more than 50 % of the total contribution of FM and more than 90% of the total contribution of the cluster. Still, overall they do not give much guidance in judging whose variables' contributions are most important in a cluster because of differences between relative contributions of individual rows and columns. □

To measure the relative influence of contributions B_{kv}, let us utilize the property that they sum up to the total data scatter and, thus, can be considered an instance of the flow data. The table of contributions can be analyzed in the same way as a contingency table (see section 2.2.3). Let us define, in particular, the relative contribution of feature v to cluster S_k, $B(k/v) = B_{kv}/T_v$, to show what part of the variable contribution goes to the cluster. The total explained part of T_v, $B_v = B_{+v}/T_v = \sum_{k=1}^{K} B(k/v)$, is equal to the correlation ratio $\eta^2(S, v)$ introduced in section 2.2.3.

More sensitive measures can be introduced to compare the relative contributions $B(k/v)$ with the contribution of cluster S_k, $B_{k+} = \sum_{v \in V} B_{kv} = N_k d(0, c_k)$, related to the total data scatter $T(Y)$. These are similar to Quetelet coefficients introduced for flow data: the difference $g(k/v) = B(k/v) - B_{k+}/T(Y)$ and the relative difference $q(k/v) = g(k/v)/(B_{k+}/T(Y)) = \frac{T(Y)B_{kv}}{T_v B_{k+}} - 1$. The former compares the contribution of v with the average contribution of variables to S_k. The latter relates this to the cluster's contribution. Index $q(k/v)$ can also be expressed as the ratio of the relative contributions of v: within S_k, B_{kv}/B_{k+}, and in the whole data, $T_v/T(Y)$. We refer to $q(k/v)$ as the Relative contribution index, $RCI(k, v)$.

For each cluster k, features v with the largest $RCI(k, v)$ should be presented to the user for interpretation.

Example 3.26. Table QScaD of the relative and Quetelet indexes

All three indexes of association, $B(k/v)$, $g(k/v)$ and $RCI\ q(k/v)$, applied to the Market towns data in Table 3.22 are presented in Table 3.23 below cluster centroids.

Table 3.23: Tendencies of the cluster structure of Market towns. At each cluster, the first and second lines show the cluster's centroid in raw and standardized scales; the other lines display the relative contribution $B(k/v)$ (Rcnt), difference $g(k/v)$ (Dcnt), and RCI $q(k, v)$, respectively, expressed as percentages. The last three lines show these three indexes applied to the explained parts of feature contributions.

k	C-d	P	PS	Do	Ho	Ba	Su	Pe	DIY	SP	PO	CAB	FM
1	Real	18484.00	7.63	3.63	1.13	11.63	4.63	4.13	1.00	1.38	6.38	1.25	0.38
	Stand	0.51	0.38	0.56	0.36	0.38	0.38	0.30	0.26	0.44	0.47	0.30	0.17
	Rcnt	58.75	51.52	54.17	28.41	50.61	44.31	29.37	30.67	36.43	57.31	17.99	3.40
	Dcnt	23.62	16.39	19.04	-6.72	15.48	9.18	-5.76	-4.46	1.30	22.18	-17.14	-31.73
	RCI	67.23	46.65	54.21	-19.12	44.05	26.14	-16.40	-12.69	3.69	63.15	-48.80	-90.31
2	Real	5267.67	2.17	0.83	0.50	4.67	1.83	1.67	0.00	0.50	1.67	0.67	1.00
	Stand	-0.10	-0.07	-0.14	0.05	0.02	-0.01	-0.05	-0.07	0.01	-0.12	0.01	0.80
	Rcnt	1.54	1.33	2.38	0.41	0.09	0.05	0.73	1.88	0.00	2,79	0.02	53.33
	Dcnt	-8.43	-8.64	-7.59	-9.57	-9.88	-9.93	-9.24	-8.09	-9.97	-7.18	-9.95	43.36
	RCI	-84.52	-86.61	-76.08	-95.93	-99.10	-99.54	-92.72	-81.17	-99.96	-72.05	-99.82	434.89
3	Real	2597.28	1.17	0.50	0.00	1.22	0.61	0.89	0.00	0.06	1.44	0.11	0.00
	Stand	-0.22	-0.15	-0.22	-0.20	-0.16	-0.19	-0.17	-0.07	-0.22	-0.15	-0.27	-0.20
	Rcnt	24.11	18.84	18.60	19.46	20.31	24.06	20.38	5.63	19.60	12.70	31.39	10.00
	Dcnt	5.51	0.24	-0.00	0.86	1.71	5.46	1.79	-12.96	1.00	-5.90	12.79	-8.60
	RCI	29.62	1.30	-0.01	4.63	9.20	29.36	9.61	-69.71	5.39	-31.70	68.78	-46.23
4	Real	11245.33	3.67	2.00	1.33	5.33	2.33	3.67	0.67	1.00	2.33	1.33	0.00
	Stand	0.18	0.05	0.16	0.47	0.05	0.06	0.23	0.15	0.26	-0.04	0.34	-0.20
	Rcnt	2.70	0.38	1.56	17.66	0.37	0.37	6.70	3.76	4.54	0.13	8.73	1.67
	Dcnt	-1.47	-3.79	-2.61	13.49	-3.80	-3.80	2.53	-0.41	0.38	-4.04	4.56	-2.50
	RCI	-35.32	-90.91	-62.62	323.75	-91.10	-91.19	60.68	-9.87	9.06	-96.94	109.47	-60.00
5	Real	5347.00	2.50	0.00	1.00	2.00	1.50	2.00	0.00	0.50	1.50	1.00	0.00
	Stand	-0.09	-0.04	-0.34	0.30	-0.12	-0.06	-0.01	-0.07	0.01	-0.14	0.18	-0.20
	Rcnt	0.48	0.17	5.09	4.86	1.26	0.29	0.00	0.63	0.00	1.28	1.55	1.11
	Dcnt	-1.12	-1.43	3.50	3.27	-0.33	-1.30	-1.59	-0.97	-1.59	-0.31	-0.04	-0.48
	RCI	-70.08	-89.58	219.92	205.73	-20.61	-81.96	-99.79	-60.66	-99.91	-19.48	-2.58	-30.17
6	Real	8674.60	3.80	2.00	0.00	3.20	2.00	2.40	0.00	0.00	2.80	0.80	0.00
	Stand	0.06	0.06	0.16	-0.20	-0.06	0.01	0.05	-0.07	-0.24	0.02	0.08	-0.20
	Rcnt	0.52	0.92	2.60	5.41	0.73	0.02	0.54	1.56	6.93	0.08	0.74	2.78
	Dcnt	-1.74	-1.34	0.34	3.15	-1.53	-2.24	-1.72	-0.69	4.67	-2.18	-1.52	0.52
	RCI	-77.04	-59.31	14.89	139.25	-67.69	-99.25	-76.27	-30.73	206.73	-96.44	-67.17	22.95
7	Real	5593.00	2.00	1.00	0.00	5.00	2.67	2.00	0.00	1.00	2.33	1.00	0.00
	Stand	-0.08	-0.09	-0.09	-0.20	0.04	0.10	-0.01	-0.07	0.26	-0.04	0.18	-0.20
	Rcnt	0.55	0.95	0.57	3.24	0.17	1.23	0.01	0.94	4.54	0.13	2.33	1.67
	Dcnt	-1.01	-0.61	-0.99	1.68	-1.39	-0.33	-1.56	-0.62	2.98	-1.43	0.76	0.11
	RCI	-64.79	-38.95	-63.22	107.78	-89.21	-20.98	-99.68	-39.84	191.16	-91.84	48.96	6.78
Ex.	Rcnt	88.64	74.11	84.97	79.45	73.54	70.32	57.72	45.07	72.05	74.42	62.74	73.96
	Dcnt	15.36	0.83	11.69	6.17	0.26	-2.95	-15.56	-28.21	-1.22	1.14	-10.54	0.68
	RCI	20.96	1.14	15.96	8.42	0.35	-4.03	-21.23	-38.49	-1.67	1.56	-14.38	0.93

Now contributions have become visible indeed. One can see, for instance, that variable Do highly contributes to cluster 5: RCI is 219.9. Why? As the upper number in the cell, 0, shows, this is a remarkable case indeed: no Doctor surgeries in the cluster at all.

The difference between clusters 6 and 7, that was virtually impossible to spot with other interpretation aids, now can be explained by the high RCI values of SP, in excess of 100%, reached at these clusters. A closer look at the data shows that there is a swimming pool in each town in cluster 7 and none in cluster 6. If the variable SP is removed then clusters 6 and 7 will not differ anymore and join together.

Overall, the seven nontrivial clusters can be considered as reflecting the following four tiers of the settlement system: largest towns (Cluster 1), small towns (Cluster

Table 3.24: ScaD for Masterpieces data in Table 3.2.

Title	LenS	LenD	NChar	SCon	Pers	Obje	Dire	Total	Total,%
Dickens	0.03	0.05	0.04	1.17	0.00	0.09	0.06	1.43	24.08
Twain	0.14	0.25	0.15	0.42	0.09	0.00	0.06	1.10	18.56
Tolstoy	0.06	0.12	0.50	0.28	0.09	0.09	0.38	1.53	25.66
Expl	0.23	0.41	0.69	1.88	0.18	0.18	0.50	4.06	68.30
Unex	0.51	0.28	0.20	0.00	0.44	0.44	0.00	1.88	31.70
Total	0.74	0.69	0.89	1.88	0.63	0.63	0.50	5.95	100.00

3), large towns (Clusters 4 and 6), and small-to-average towns (Clusters 2,5 and 7). In particular, the largest town Cluster 1 consists of towns whose population is two to three times larger than the average, and they have respectively larger numbers of all facilities, of which even more represented are Post Offices, Doctors, Primary Schools, and Banks. The small town Cluster 3 consists of the smallest towns with 2-3 thousand residents on average. Respectively, the other facilities are also smaller and some are absent altogether (such as DIY shops and Farmers' markets). Two large town clusters, Cluster 4 and Cluster 6, are formed by towns of nine to twelve thousand residents. Although lack of such facilities as Farmers' market is common to them, Cluster 4 is by far the richer, with service facilities that are absent in Cluster 6, which probably is the cause of the separation of the latter within the tier. Three small-to-average town clusters have towns of about 5,000 residents and differ from each other by the presence of a few fancy objects that are absent from the small town cluster, as well as from the other two clusters of this tier. These objects are: a Farmers' market in Cluster 2, a Hospital in Cluster 5, and a Swimming pool in Cluster 7. □

Example 3.27. ScaD and QScaD for Masterpieces

Tables 3.24 and 3.25 present similar decompositions with respect to author-based clustering of the Masterpieces data in Table 2.13 on page 61. This time, only Quetelet indexes of variables, $RCI(k, v)$ are presented (in Table 3.25).

Table 3.25 shows feature SCon as the one most contributing to the Dickens cluster, feature LenD to the Twain cluster, and features NChar and Direct to the Tolstoy cluster. Indeed, these clusters can be distinctively described by the statements "SCon=0," "LenD $<$ 28," and "NChar $>$ 3" (or "Narrative is Direct"), respectively. Curiously, the decisive role of LenD for the Twain cluster cannot be recognized from the absolute contributions in Table 3.24: SCon prevails over the Twain cluster in that table. □

Table 3.25: Relative centroids: cluster centroids standardized and Relative contribution indexes of variables, in cluster first and second lines, respectively.

Title	LenS	LenD	NChar	SCon	Pers	Obje	Dire
Dickens	0.10	0.12	-0.11	-0.63	0.17	-0.02	-0.14
	-83.3	-70.5	-81.5	**158.1**	-40.1	-100.0	-50.5
Twain	-0.21	-0.29	-0.22	0.38	-0.02	0.17	-0.14
	1.2	**91.1**	-9.8	20.2	-100.0	-22.3	-35.8
Tolstoy	0.18	0.24	0.50	0.38	-0.22	-0.22	0.43
	-68.3	-33.0	**119.5**	-41.5	-43.3	-43.3	**197.0**

3.4.3 Cluster representatives

The user can be interested in a conceptual description of a cluster, but he also can be interested in looking at the cluster via its representative, a "prototype." This is especially appealing when the representative is a well known object. Such an object can give much better meaning to a cluster than a logical description in situations where entities are complex and the concepts used in description are superficial and do not penetrate deep into the phenomenon. This is the case, for instance, in mineralogy where a class of minerals can be represented by its stratotype, or in literary studies where a general concept can be represented by a literary character.

To specify what entity should be taken as a representative of its cluster, conventionally that entity is selected which is the nearest to its cluster's centroid. This strategy can be referred to as "the nearest in distance." It can be justified in terms of the square error criterion $W(S,c) = \sum_{k=1}^{K} \sum_{h \in S_k} d(y_h, c_k)$ (3.2). Indeed, the entity $h \in S_k$ which is the nearest to c_k contributes the least to $W(S,c)$, that is, to the unexplained part of the data scatter.

The contribution based approach supplements the conventional approach. Decomposition of the data scatter (3.4) suggests a different strategy by relating to the explained rather than unexplained part of the data scatter. This strategy suggests that the cluster's representative must be the entity that maximally contributes to the explained part, $B(S,c) = \sum_{k=1}^{K} \sum_v c_{kv}^2 N_k$.

How can one compute the contribution of an entity to that? There seems nothing of entities in $B(S,c)$. To reveal contributions of individual entities, let us recall that $c_{kv} = \sum_{i \in S_k} y_{iv}/N_k$. Let us take c_{kv}^2 in $B(S,c)$ as the product of c_{kv} with itself, and change one of the factors for the definition. This way we obtain equation $c_{kv}^2 N_k = \sum_{i \in S_k} y_{iv} c_{kv}$. This leads to a formula for $B(S,c)$ as the summary inner product:

$$B(S,c) = \sum_{k=1}^{K} \sum_{i \in S_k} \sum_{v \in V} y_{iv} c_{kv} = \sum_{k=1}^{K} \sum_{i \in S_k} (y_i, c_k), \tag{3.8}$$

which shows that the contribution of entity $i \in S_k$ is (y_i, c_k).

The most contributing entity is "the nearest in inner product" to the cluster centroid, which may lead sometimes to different choices. Intuitively, the choice according to the inner product follows tendencies represented in c_k towards the whole of the data rather than c_k itself, which is manifested in the choice according to distance.

Example 3.28. Different concepts of cluster representatives

The entity based elements of the data scatter decomposition for the Dickens cluster from Table 3.24 are displayed in Table 3.26. Now some contributions are negative, which shows that a feature at an entity may be at odds with the cluster centroid. According to this table, the maximum contribution to the data scatter, 8.82%, is

Table 3.26: Decomposition of feature contributions to the Dickens cluster in Table 3.24 (in thousandth). The right-hand column shows distances to the cluster's centroid.

Title	LenS	LenD	NChar	SCon	Pers	Obje	Dire	Cntr	Cntr,%	Dist
OTwist	-19	29	37	391	61	5	21	524	8.82	222
DoSon	38	6	0	391	61	5	21	521	8.77	186
GExpect	8	12	0	391	-36	-9	21	386	6.49	310
Dickens	27	46	37	1172	86	2	62	1431	24.08	0

Table 3.27: Two Dickens' masterpieces along with features contributing to their differences.

Item	LenS	LenD	NChar
OTwist	19.0	43.7	2
DoSon	29.4	36.0	3
Cluster mean	24.1	39.2	2.67
Grand mean	22.4	34.1	3.00

delivered by the novel Oliver Twist. Yet the minimum distance to the cluster's centroid is reached at a different novel, Dombey and Son.

To see why this may happen, let us take a closer look at the two novels versus within cluster and grand means (Table 3.27).

Table 3.27 clearly shows that the cluster's centroid is greater than the grand mean on the first two components and smaller on the third one. These tendencies are better expressed in Dombey and Son over the first component and in Oliver Twist over the other two, which accords with the contributions in Table 3.26. Thus, Oliver Twist wins over Dombey and Son as better representing the differences between the cluster centroid and the overall gravity center, expressed in the grand mean. With the distance measure, no overall type tendency can be taken into account. □

Example 3.29. Interpreting Bribery clusters

Let us apply similar considerations to the five clusters of the Bribery data listed in Table 3.16. Since individual cases are not of interest here, no cluster representatives will be considered. However, it is highly advisable to consult the original data and their description on page 19.

In cluster 1, the most contributing features are: Other branch (777%), Change of category (339%), and Level of client (142%). Here and further in this example the values in parentheses are relative contribution indexes RCI. By looking at the cluster's centroid, one can find specifics of these features in the cluster. In particular, all its cases appear to fall in Other branch, comprising such bodies as universities or hospitals. In each of the cases the client's issue was of a personal matter, and most times (six of the eight cases) the service provided was based on re-categorization of the client into a better category. The category Other branch (of feature Branch) appears to be distinctively describing the cluster: the eight cases in this category constitute the cluster.

Cluster 2 consists of nineteen cases. Its most salient features are: Obstruction of justice (367%), Law enforcement (279%), and Occasional event (151%). By looking at the centroid values of these features, one can conclude: (1) all corruption cases in this cluster have occurred in the law enforcement system; (2) they are mostly done via obstruction of justice for occasional events. The fact (1) is not sufficient for distinctively describing the cluster since there are thirty-four cases, not just nineteen, that have occurred in the law enforcement branch. Two more conditions have been found by a cluster description algorithm, APPCOD (see in section 7.1), to be conjunctively added to (1) to make the description distinctive: (3) the cases occurred at office levels higher than Organization, and (4) no cover-up was involved.

Cluster 3 contains ten cases for which the most salient categories are: Extortion in variable III Type of service (374%), Organization (189%), and Government (175%) in X Branch. Nine of the ten cases occurred in the Government branch, overwhelmingly at the level of organization (feature I) and, also overwhelmingly, the office workers extorted money for rendering their supposedly free services (feature III). The client level here is always of an organization, though this feature is not that salient as the other three features.

Cluster 4 contains seven cases, and its salient categories are: Favors in III (813%), Government in X (291%), and Federal level of Office (238%). Indeed, all its cases occurred in the government legislative and executive branches. The service provided was mostly Favors (six of seven cases). Federal level of corrupt office was not frequent, two cases only. Still, this frequency was much higher than the average, for the two cases are just half of the total number, four, of the cases in which Federal level of office was involved.

Cluster 5 contains eleven cases and pertains to two salient features: Cover-up (707%) and Inspection (369%). All of the cases involve Cover-up as the service provided, mostly in inspection and monitoring activities (nine cases of eleven). A distinctive description of this cluster can be defined to conjunct two statements: it is always a cover-up but not at the level of Organization.

Overall, the cluster structure leads to the following overview of the situation. Most important, it is Branch which is the feature defining Russian corruption when looked at through the media glass. Different branches tend to involve different corruption services. The government corruption involves either Extortion for rendering their free services to organizations (Cluster 3) or Favors (Cluster 4). The law enforcement corruption in higher offices is for either Obstruction of justice (Cluster 2) or Cover-up (Cluster 5). Actually, Cover-up does not exclusively belong in the law enforcement branch: it relates to all offices that are to inspect and monitor business activities (Cluster 5). Corruption cases in Other branch involve re-categorization of individual cases into more suitable categories (Cluster 1). □

3.4.4 Measures of association from ScaD tables

Here we are going to see that summary contributions of clustering towards a feature in ScaD tables are compatible with traditional statistical measures of correlation considered in section 2.2.

Quantitative feature case: Correlation ratio

As proven in section 5.2.3, the total contribution $B_{+v} = \sum_k B_{vk}$ of a quantitative feature v to the cluster-explained part of the scatter, presented in the ScaD tables, is proportional to the correlation ratio between v and cluster partition S, introduced in section 2.2.2. In fact, the correlation ratios can be found by relating the row Expl to row Total in the general ScaD table 3.21.

Example 3.30. Correlation ratio from a ScaD table
The correlation ratio of the variable P (Population resident) over the clustering in Table 3.22 can be found by relating the corresponding entries in rows Expl and Total; it is 3.16/3.56=0.89. This relatively high value shows that the clustering closely – though not entirely – follows this variable. In contrast, the clustering has rather little to do with variable DIY, the correlation ratio of which is equal to 0.79/1.75=0.45. □

Categorical feature case: Chi-square and other contingency coefficients

The summary contribution of a nominal feature l having V_l as the set of its categories, to the clustering partition S has something to do with contingency coefficients introduced in section 2.2.3. It is proven in section 5.2.4 to be equal to

$$B(S,l) = \frac{N}{|V_l|}\sum_{k=1}^{K}\sum_{v\in V_l}\frac{(p_{kv}-p_{k+}p_{+v})^2}{p_{k+}b_v^2} \quad (3.9)$$

where b_v stands for the scaling coefficient at the data standardization. Divisor $|V_l|$, the number of categories, comes from the rescaling stage introduced in section 2.4.

The coefficient $B(S,l)$ in (3.9) can be further specified depending on the scaling coefficients b_v. In particular, the items summed up in (3.9) are:

1. $\frac{(p_{kv}-p_k p_v)^2}{p_k}$ if $b_v = 1$, the range;
2. $\frac{(p_{kv}-p_k p_v)^2}{p_k p_v(1-p_v)}$ if $b_v = \sqrt{p_v(1-p_v)}$, the Bernoullian standard deviation;
3. $\frac{(p_{kv}-p_k p_v)^2}{p_k p_v}$ if $b_u = \sqrt{p_u}$, the Poissonian standard deviation.

Items 1 and 3 above lead to $B(S,l)$ being equal to the summary Quetelet coefficients introduced in section 2.2.3. The Quetelet coefficients, thus, appear to be related to the data standardization. Specifically, G^2 corresponds to $b_v = 1$ and $Q^2 = X^2$ to $b_v = \sqrt{p_v}$. Yet item 2, the Bernoullian standardization, leads to an association coefficient which has not been considered in the literature.

Example 3.31. ScaD based association between a feature and clustering
Let us consider the contingency table between the author-based clustering of masterpieces and the only nominal variable in the data, Narrative (Table 3.28). In this example, the dummy variables have been range normalized and then rescaled with $b'_v = \sqrt{3}$, which is consistent with formula (3.9) with $b_v = 1$ and $|V_l| = 3$ for the calculation of the summary contribution $B(S,l)$. Table 3.29 presents the values of $\frac{(p_{kv}-p_k p_v)^2}{3p_k/N}$ in each cell of the cross classification. In fact, these are entries of the full ScaD table in Table 3.24, page 104, related to the categories of Narrative (columns) and the author-based clusters (rows), with row Total corresponding to row Expl in Table 3.24. In particular, the total contribution of the clustering and variable Narrative is equal to 0.18+0.18+0.50=0.86, or about 14.5% of the data scatter. □

Table 3.28: Cross-classification of the author-based partition and Narrative at the eight masterpieces (in thousandth).

Class	Personal	Objective	Direct	Total
Dickens	125	250	0	375
Twain	250	125	0	375
Tolstoy	0	0	250	250
Total	375	375	250	1000

Table 3.29: Elements of calculation $B(S, l)$ according to formula (3.9) (in ten-thousandth).

Class	Personal	Objective	Direct	Total
Dickens	17	851	625	1493
Twain	851	17	625	1493
Tolstoy	938	938	3750	5626
Total	1806	1806	5000	8606

3.5 Overall assessment

K-Means advantages:

1. Models typology building activity.
2. Computationally effective both in memory and time.
3. Can be utilized incrementally, "on-line."
4. Straightforwardly associates feature salience weights with feature scales.
5. Applicable to both quantitative and categorical data and mixed data provided that care has been taken of the relative feature scaling.
6. Provides a number of interpretation aids including cluster prototypes and features and entities most contributing to cluster specificity.

K-Means issues:

1. Simple convex spherical shape of clusters.
2. Choosing the number of clusters and initial seeds.
3. Instability of results with respect to initial seeds.

The issues above are not necessarily shortcomings. To cope with issue 1, the feature set should be chosen carefully according to the goals of the data analysis. To cope with issue 2, the initial seeds should be selected based on conceptual understanding of the substantive domain or preliminary data analysis with the AP clustering approach. There can be some advantages in the issues as well. Issue 3 keeps solutions close to pre-specified centroid settings, which is good when centroids have been conceptually substantiated. Issue 1 of simplicity of cluster shapes provides for a possibility of deriving simple conjunctive descriptions of the clusters, which can be used as supplementary interpretation aids (see section 6.3).

A clustering algorithm should present the user with a comfortable set of options to do clustering. In our view, the intelligent version of K-Means described above and its versions, implementing the possibility of removal of entities that have been found either (1) "deviant" (contents of small Anomalous pattern clusters), or (2) "intermediate" (entities that are far away from their centroids, or have small attraction index values), or (3) "trivial" (entities that are close to the grand mean), give the user an opportunity to select a preferred option without imposing on him technical issues.

Chapter 4

Ward Hierarchical Clustering

After reading this chapter the reader will know about:

1. Agglomerative and divisive clustering.
2. The Ward algorithm for agglomerative clustering.
3. Divisive algorithms for Ward criterion.
4. Visualization of hierarchical clusters with heighted tree diagrams and box charts.
5. Decomposition of the data scatter involving both Ward and K-Means criteria.
6. Contributions of individual splits to: (i) the data scatter, (ii) feature variances and covariances, and (iii) individual entries.
7. Extensions of Ward clustering to dissimilarity, similarity and contingency data.

Base words

Agglomerative clustering Any method of hierarchical clustering that works bottom up, by merging two nearest clusters at each step.

Aggregation Transformation of a contingency table into a smaller size table by aggregating its row and column categories with summing up corresponding entries. Can be done with Ward clustering extended to contingency tables.

Box chart A visual representation of an upper cluster hierarchy involving a triple partition of a rectangular box corresponding to each split. The middle part is proportional to the contribution of the split and the other two to contributions of resulting clusters.

Conceptual clustering Any divisive clustering method that uses only a single feature at each splitting step. The purity and category utility scoring functions are closely related to Ward clustering criterion.

Contribution of a split Part of the data scatter that is explained by a split and equal to the Ward distance between split parts. Features most contributing to a split can be used in taxonomic analysis. Split contributions to covariances between features and individual entities can also be considered.

Divisive clustering Any method of hierarchical clustering that works from top to bottom, by splitting a cluster in two distant parts, starting from the universal cluster containing all entities.

Heighted tree A visual representation of a cluster hierarchy by a tree diagram in which nodes correspond to clusters and are positioned along a vertical axis in such a way that the height of a parent node is always greater than the heights of its child nodes.

Hierarchical clustering An approach to clustering based on representation of data as a hierarchy of clusters nested over set-theoretic inclusion. In most cases, hierarchical clustering is used as a tool for partitioning, though there are some cases, such as that of the evolutionary tree, in which the hierarchy reflects the substance of a phenomenon.

Ward clustering A method of hierarchical clustering involving Ward distance between clusters. Ward distance is maximized in Ward divisive clustering and minimized in Ward agglomerative clustering. Ward clustering accords with the data recovery approach.

Ward distance A measure of dissimilarity between clusters, equal to the squared Euclidean distance between cluster centroids weighted by the product of cluster sizes.

4.1 Agglomeration: Ward algorithm

Hierarchical clustering is a discipline devoted to presenting data in the form of a hierarchy over the entity set. Sometimes features are also put into the same or separate hierarchy. There are two approaches to building a cluster hierarchy:

(a) agglomerative clustering that builds a hierarchy in the bottom-up fashion by starting from smaller clusters and sequentially merging them into 'parental' nodes, and

(b) divisive clustering that builds a hierarchy top-to-bottom by splitting greater clusters into smaller ones starting from the entire data set.

The agglomerative approach in clustering builds a cluster hierarchy by merging two clusters at a time, starting from singletons (one-entity clusters) or other pre-drawn clusters. Thus, each non-singleton cluster in the hierarchy is the union of two smaller clusters, which can be drawn like a genealogical tree. The smaller clusters are called children of the united cluster which is referred to as their parent. The singletons are referred to as terminal nodes or leaves, and the universal combined cluster consisting of the entire entity set is referred to as the root of the tree. All other clusters are referred to as nodes of the tree.

The singletons and their successive mergers at every intermediate step form what is called a lower cluster hierarchy, until the root is reached, at which point a full cluster hierarchy emerges.

Besides the tree topology, some metric information is usually also specified: each cluster-node is accompanied with a positive number referred to as its height. A heighted tree is drawn in such a way that each cluster is represented with a node whose height is reflected in its position over the vertical axis. The heights then should satisfy the natural monotonicity requirement: the parent's height is greater than its children's heights.

At each step of an agglomerative clustering algorithm a set of already formed clusters $\mathbf{S}$ is considered along with the matrix of distances between maximal clusters. Then two nearest maximal clusters are merged and the newly formed cluster is supplied with its height and distances to other clusters. The process ends, typically, when all clusters have been merged into the universal cluster consisting of the set I of all entities under consideration.

Agglomerative algorithms differ depending on between-cluster distance measures used in them. Especially popular are the so-called single linkage, full linkage and group average criteria. The distance between clusters is defined as the minimum or maximum distance between cluster elements in the single linkage and full linkage methods, respectively. The group average criterion takes the distance between cluster centroids as the between-cluster distance. Quite a broad set of agglomerative algorithms has been defined by Lance and Williams in terms of a formula for dynamic recalculation of the distances between clusters being merged and other clusters into the distances from the merged cluster (see, for instance, [75, 58, 90]). Lance and Williams formula covers all interesting algorithms proposed in the literature so far and much more.

Here we concentrate on a weighted group average criterion first proposed by Ward [135].

Specifically, for clusters S_{w1} and S_{w2} whose cardinalities are N_{w1} and N_{w2} and centroids c_{w1} and c_{w2}, respectively, Ward distance is defined as

$$dw(S_{w1}, S_{w2}) = \frac{N_{w1}N_{w2}}{N_{w1} + N_{w2}} d(c_{w1}, c_{w2}) \tag{4.1}$$

where $d(c_{w1}, c_{w2})$ is the squared Euclidean distance between c_{w1} and c_{w2}.

To describe the intuition behind this criterion, let us consider a partition S on I and two of its classes S_{w1} and S_{w2} and ask ourselves the following question: how the square error of S, $W(S, c)$, would change if these two classes are merged together?

To answer the question, let us consider the partition that differs from S only in that respect that classes S_{w1} and S_{w2} are changed in it for the union $S_{w1} \cup S_{w2}$ and denote it by $S(w1, w2)$. Note that the combined cluster's centroid can be expressed through centroids of the original classes as $c_{w1 \cup w2} = (N_{w1}c_{w1} + N_{w2}c_{w2})/(N_{w1} + N_{w2})$. Then calculate the difference between the square error criterion values at the two partitions, $W(S(w1, w2), c(w1, w2)) - W(S, c)$, where $c(w1, w2)$ stands for the set of centroids in $S(w1, w2)$. The difference is equal to the Ward distance between S_{w1} and S_{w2}:

$$dw(S_{w1}, S_{w2}) = W(S(w1, w2), c(w1, w2)) - W(S, c) \tag{4.2}$$

Because of the additive nature of the square error criterion (3.2), all items on the right of (4.2) are self subtracted except for those related to S_{w1} and S_{w2} so that the following equation holds

$$dw(S_{w1}, S_{w2}) = W(S_{w1} \cup S_{w2}, c_{w1 \cup w2}) - W(S_{w1}, c_{w1}) - W(S_{w2}, c_{w2}) \tag{4.3}$$

where, for any cluster S_k with centroid c_k, $W(S_k, c_k)$ is the summary distance (3.1) from elements of S_k to c_k ($k = w1, w2, w1 \cup w2$).

The latter equation can be rewritten as

$$W(S_{w1} \cup S_{w2}, c_{w1 \cup w2}) = W(S_{w1}, c_{w1}) + W(S_{w2}, c_{w2}) + dw(S_{w1}, S_{w2}), \tag{4.4}$$

which shows that the summary square error of the merged cluster is the sum of square errors of the original clusters and Ward distance between them.

Since all expressions on the right side in (4.4) are positive, the square error $W(S_{w1} \cup S_{w2}, c_{w1 \cup w2})$ of the merged cluster is always greater than that of either of the constituent clusters, which allows using the cluster square error as the height function in visualizing a cluster hierarchy.

Equation (4.2) justifies the use of Ward distance if one wants to keep the within cluster variance as small as possible at each of the agglomerative steps. The following presents the Ward agglomeration algorithm.

Ward algorithm:
1. *Initial setting.* The set of maximal clusters is all the singletons, their cardinalities being unity, heights zero, themselves being centroids.
2. *Cluster update.* Two clusters, S_{w1} and S_{w2}, that are closest to each other (being at the minimum Ward distance) among the maximal clusters, are merged together forming their parent cluster $S_{w1\cup w2} = S_{w1} \cup S_{w2}$. The merged cluster's cardinality is defined as $N_{w1\cup w2} = N_{w1} + N_{w2}$, centroid as $c_{w1\cup w2} = (N_{w1}c_{w1} + N_{w2}c_{w2})/N_{w1\cup w2}$ and its height as $h(w1 \cup w2) = h(w1) + h(w2) + dw(S_{w1}, S_{w2})$.
3. *Distance update.* Put $S_{w1\cup w2}$ into and remove S_{w1} and S_{w2} from the set of maximal clusters. Define Ward distances between the new cluster $S_{w1\cup w2}$ and other maximal clusters S_t.
4. *Repeat.* If the number of maximal clusters is larger than 1, go to step 2. Otherwise, output the cluster merger tree along with leaves labelled by the entities.

Ward agglomeration starts with singletons whose variance is zero and produces an increase in criterion (3.2) that is as small as possible, at each agglomeration step. This justifies the use of Ward agglomeration results by practitioners to get a reasonable initial setting for K-Means. Two methods supplement each other in that clusters are carefully built with Ward agglomeration, and K-Means allows overcoming the inflexibility of the agglomeration process over individual entities by reshuffling them. There is an issue with this strategy though: Ward agglomeration, unlike K-Means, is a computationally intensive method, not applicable to large sets of entities.

The height of the new cluster is defined as its square error according to equation (4.4). Since the heights of merged clusters include the sums of heights of their children, the heights of the nodes grow fast, with an "exponential" speed. This can be used to address the issue of determining what number of clusters is "relevant" to the data by cutting the hierarchical tree at the layer separating long edges from shorter ones if such a layer exists (see, for example, Figure 4.1 whose three tight clusters can be seen as hanging on longer edges), see [28], pp. 76-77, for heuristical rules on this matter.

Table 4.1: Matrix of Ward distances between eight entities in Table 3.2.

Entity	1	2	3	4	5	6	7	8
1	0.00	0.25	0.44	0.57	1.10	1.13	1.15	1.51
2	0.25	0.00	0.39	0.78	0.91	1.50	0.95	1.21
3	0.44	0.39	0.00	0.97	0.58	0.92	0.91	1.19
4	0.57	0.78	0.97	0.00	0.49	0.44	0.61	1.23
5	1.10	0.91	0.58	0.49	0.00	0.37	0.41	0.94
6	1.13	1.50	0.92	0.44	0.37	0.00	0.84	1.71
7	1.15	0.95	0.91	0.61	0.41	0.84	0.00	0.30
8	1.51	1.21	1.19	1.23	0.94	1.71	0.30	0.00

Example 4.32. Agglomerative clustering of Masterpieces

Let as apply the Ward algorithm to the pre-processed and standardized Masterpieces data in Table 2.13 presented in the right-bottom display of Figure 2.5. The algorithm starts with the matrix of Ward distances between all singletons, that is a matrix of entity-to-entity Euclidean distances squared and divided by two, as obviously follows from (4.1). The Ward distance matrix is presented in Table 4.1.

Minimum non-diagonal value in the matrix of Table 4.1 is $dw(1,2) = 0.25$ with $dw(7,8) = 0.30$ and $d(5,6) = 0.37$ as the second and third runners-up, respectively. These are the starting agglomerations according to Ward algorithm: clusters $\{1,2\}$ $\{7,8\}$ and $\{5,6\}$ whose heights are 0.25, 0.30 and 0.37, respectively, shown on Figure 4.1 as percentages of the data scatter $T(Y) = \sum_{i,v} y_{iv}^2$ which is the height of the maximum cluster comprising all the entities as proven in section 5.3. Further mergers are also shown in Figure 4.1 with their heights. The author based classes hold on

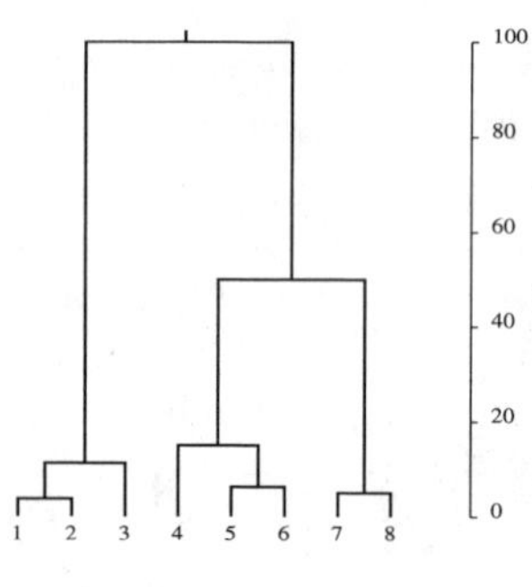

Figure 4.1: Cluster tree built with Ward clustering algorithm; the node heights are scaled in per cent to the height of the entire entity set.

the tree for about 35% of its height, then the Leo Tolstoy cluster merges with that of Mark Twain as should be expected from the bottom-right display in Figure 2.5. The hierarchy drastically changes if a different feature scaling system is applied. For instance, with the standard deviation based standardization, Leo Tolstoy's two novels do not constitute a single cluster but are separately merged within the Dickens and Twain clusters. This does not change even with the follow-up rescaling of categories of Narrative by dividing them over $\sqrt{3}$. □

4.2 Divisive clustering with Ward criterion

A divisive algorithm builds a cluster hierarchy from top to bottom, each time by splitting a cluster in two, starting from the entire set. Such an algorithm will be referred to as a Ward-like divisive clustering algorithm if the splitting steps maximize Ward distance between split parts. Let us denote a cluster by S_w, its split parts by S_{w1} and S_{w2}, so that $S_w = S_{w1} \cup S_{w2}$ and consider equation (4.4) which is applicable here: it decomposes the square error, that is, the summary distance between elements and the centroid of S_w into the sum of the square errors of the split parts and the Ward distance between them:

$$W(S_w, c_w) = W(S_{w1}, c_{w1}) + W(S_{w2}, c_{w2}) + dw(c_{w1}, c_{w2}) \tag{4.5}$$

where the indexed c refers to the centroid of the corresponding cluster.

In the process of divisions, a divisive clustering algorithm builds what is referred to as an upper cluster hierarchy, which is a binary tree rooted at the universal cluster I such that its leaves are not necessarily singletons. One may think of an upper cluster hierarchy as a cluster hierarchy halfway through construction from top to bottom, in contrast to lower cluster hierarchies that are cluster hierarchies built halfway through, bottom up.

Thus, a Ward-like divisive clustering algorithm goes like this.

Ward-like divisive clustering
1. *Start.* Put $S_w \leftarrow I$ and draw the tree root as a node corresponding to S_w at the height of $W(S_w, c_w)$.
2. *Splitting.* Split S_w in two parts, S_{w1} and S_{w2}, to maximize Ward distance $wd(S_{w1}, S_{w2})$.
3. *Drawing attributes.* In the drawing, add two children nodes corresponding to S_{w1} and S_{w2} at the parent node corresponding to S_w, their heights being their square errors.
4. *Cluster set's update.* Set $S_w \leftarrow S_{w'}$ where $S_{w'}$ is the node of maximum height among the leaves of the current upper cluster hierarchy.
5. *Halt.* Check the stopping condition (see below). If it holds, halt and output the hierarchy and interpretation aids described in section 4.2.3; otherwise, go to 2.

The following can be used as the stopping condition for the process of divisions:

1. **The number of final clusters.** The number of terminal nodes (leaves) has reached a pre-specified threshold.

2. **Cluster height.** The height $W(S_w, c_w)$ of S_w has decreased to a pre-specified threshold such as the average contribution of a single entity, $T(Y)/N$, or a pre-specified proportion, say 5%, of the data scatter.
3. **Contribution to data scatter.** The total contribution of the current cluster hierarchy, that is, the sum of Ward distances between split parts in it, has reached a pre-specified threshold such as 50% of the data scatter.

Each of these effectively specifies the number of final clusters. Other approaches to choosing the number of clusters are reviewed in section 7.5.1.

It should be noted that the drawn representation of an upper cluster hierarchy may follow formats differing from that utilized for representing results of an agglomerative method. In particular, we suggest that one can utilize the property that all contributions are summed up to 100% of the data scatter and present the process of divisions with a box chart such as in Figure 4.3 on page 122. At such a box chart each splitting is presented with a partition of a corresponding box in three parts of which that in the middle corresponds to the split whereas those on the right and left correspond to split clusters. The parts' areas are proportional to their contributions, to that of the split, Ward distance itself, and those of clusters split, which are the summary distances of the cluster's entities to their centroids, $W(S_w, c_w) = \sum_{i \in S_w} d(y_i, c_w)$, for any $S_w \in \mathbf{S}$. The box chart concept is similar to that of the pie chart except for the fact that the pie chart slices are of the same type whereas there are two types of slices in the box chart, those corresponding to splits and those to split clusters.

4.2.1 2-Means splitting

Developing a good splitting algorithm at Step 2 of Ward-like divisive clustering is an issue. To address it, let us take a closer look at the Ward distance as a splitting criterion. One of the possibilities follows from the fact that maximizing Ward distance is equivalent to minimizing the square-error criterion $W(S, c)$ of K-Means at $K = 2$ as proven in section 5.3.3 [90]. Thus, 2-Means can be used in the Ward-like divisive clustering algorithm to specify it as follows.

2-Means splitting in divisive clustering
1. *Initial setting.* Given $S_w \subseteq I$, specify initial seeds of split parts, $c_{w1} = y_1$ and $c_{w2} = y_2$.
2. *Straight 2-Means.* Apply 2-Means algorithm to S_w with initial seeds specified at step 1 and the Euclidean distance squared (2.17).
3. *Output* results: (a) split parts S_{w1} and S_{w2}; (b) their centroids c_{w1} and c_{w2} along with their heights, $h(S_1)$ and $h(S_2)$; (c) contribution of the split, that is, Ward distance between S_{w1} and S_{w2}.

In spite of the fact that Euclidean squared distance d, not Ward distance dw is used in splitting, the algorithm in fact goes in line with Ward agglomeration.

To specify two initial seeds in 2-Means splitting, either of the three options indicated in section 3.2 can be applied:

1. random selection;

2. maximally distant entities;

3. centroids of two Anomalous pattern clusters derived at S_w.

Random selection must be repeated many times to get a reasonable solution for any sizeable data. The 2-Means splitting algorithm has two major drawbacks:

1. Step 1 is highly time consuming since it requires finding the maximum of all pair-wise distances.

2. The result can be highly affected by the choice of the initial seeds as the most distant entities, which can be at odds with the cluster structure hidden in data.

In spite of these, divisive clustering with 2-Means became rather popular after it had been experimentally approved in [127], where it was described as a heuristical method under the name of "Bisecting K-Means" probably without any knowledge of the work [90] in which it was proposed as an implementation of divisive clustering with the Ward criterion.

4.2.2 Splitting by separating

To relax both of the issues above, one can employ a different formulation of Ward distance in (4.1):

$$dw(S_{w1}, S_{w2}) = \frac{N_w N_{w1}}{N_{w2}} d(c_{w1}, c_w) \tag{4.6}$$

in which the center of S_{w2} is changed for the center of the parent cluster S_w along with changing the numeric factor in the distance. Expression (4.6) expresses Ward distance through one of the split parts only. The proof easily follows from equation $N_w c_w = N_{w1} c_{w1} + N_{w2} c_{w2}$ and the definition of the squared Euclidean distance d.

To maximize (4.6), one needs to keep track of just one cluster S_{w1} because c_w and N_w are pre-specified by S_w and do not depend on splitting.

We leave the task of reformulation of the Straight 2-Means-like splitiing algorithm with criterion (4.6) to the reader. Instead, we produce a version exploiting the incremental approach to building clusters described in section 3.1.3. According to this approach, cluster S_{w2} and its center c_{w2} are updated incrementally by considering one entity's move at a time. Let us denote $z = 1$ if an entity was added to S_{w1} and $z = -1$ if that entity was removed. Then the new value of Ward distance (4.6) after the move will be $N_w(N_{w1} + z)d(c'_{w1}, c_w)/(N_{w2} - z)$ where c'_{w1} is the updated centroid of S_{w1}. Relating this to dw (4.6), we can see that the value of Ward distance increases if the ratio is greater than 1, that is, if

$$\frac{d(c_w, c_{w1})}{d(c_w, c'_{w1})} < \frac{N_{w1}N_{w2} + zN_{w2}}{N_{w1}N_{w2} - zN_{w1}} \tag{4.7}$$

and it decreases otherwise. This leads us to the following incremental splitting algorithm.

Splitting by separating
1. *Initial setting.* Given $S_w \subseteq I$ and its centroid c_w, specify its split part S_{w1} as consisting of entity y_1, which is furthest from c_w, and put $c_{w1} = y_1$, $N_{w1} = 1$ and $N_{w2} = N_w - 1$.
2. *Next move.* Take an entity y_i; this can be that nearest to c_{w1}.
3. *Stop-condition.* Check inequality (4.7) with y_i added to S_{w1} if $y_i \notin S_{w1}$ or removed from S_{w1}, otherwise. If (4.7) holds, change the state of y_i with respect to S_{w1} accordingly, recalculate $c_{w1} = c'_{w1}$, N_{w1}, N_{w2} and go to step 2.
4. *Output* results: split parts S_{w1} and $S_{w2} = S_w - S_{w1}$; their centroids c_{w1} and c_{w2}; their heights, $h_1 = W(S_{w1}, c_{w1})$ and $h_2 = W(S_{w2}, c_{w2})$; and the contribution of the split, that is, Ward distance $dw(S_{w1}, S_{w2})$.

To specify the seed at Step 1, the entity which is the farthest from the centroid is taken. However, different strategies can be pursued too: (a) random selection or (b) taking the centroid of the Anomalous pattern found with the AP algorithm from section 3.2.3. These strategies are similar to those suggested for the 2-Means splitting algorithm, and so are their properties.

Example 4.33. Divisive clustering of Masterpieces with 2-Means splitting
Let us apply the Ward-like divisive clustering method to the Masterpieces data in Table 2.13 range standardized with the follow-up rescaling the dummy variables corresponding to the three categories of Narrative. The method with 2-Means splitting may produce a rather poorly resolved picture if the most distant entities, 6 and 8 according to the distance matrix in Table 4.1, are taken as the initial seeds. Then step 2 would produce tentative classes $\{1, 3, 4, 5, 6\}$ and $\{2, 7, 8\}$ because 2 is closer to 8 than to 6 as easily seen in Table 4.1. This partition breaks the authorship clusters.

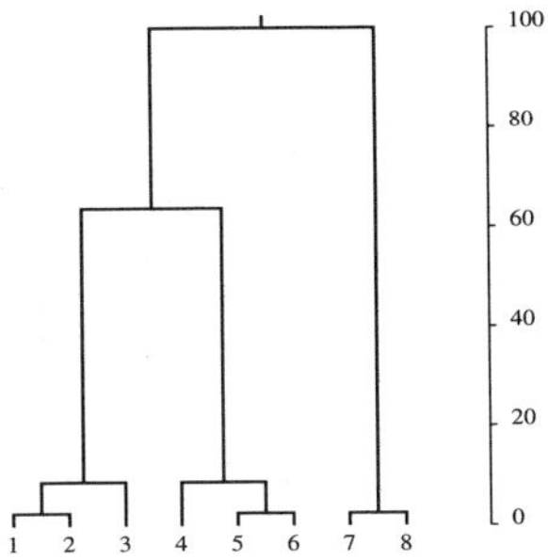

Figure 4.2: Cluster tree of masterpieces built with Splitting by separating; the node heights are scaled as percentages to the pre-processed data scatter.

Unfortunately, no further iterations can change it. This shows how vulnerable results found with the rule for initial seed setting at the two farthest entities can be.

□

Example 4.34. Divisive clustering of Masterpieces with splitting by separating and a box-chart

Splitting with separating by taking the first seed at the entity 8 which is the farthest from the origin, works more gently and produces the tree presented in Figure 4.2. This tree differs from the tree found with the agglomerative Ward method not only in the order of author-based divisions (the Tolstoy cluster first goes here) but also in the node heights.

Let us illustrate the splitting process in the Ward-like divisive clustering with a box chart. The first split separates Tolstoy's two novels, 7 and 8, from the rest. Contributions are calculated according to the decomposition (4.5). The split itself contributes to the data scatter 34.2%, the Tolstoy cluster 5.1% and the rest 60.7%, which is reflected in the areas occupied by the vertically split parts in Figure 4.3. The second split (horizontal lines across the right-hold part of the box) produces the Dickens cluster, with entities 1, 2, and 3, contributing 12.1% to the data scatter, and the Twain cluster, with entities 4, 5, and 6, contributing 14.5%; the split itself contributes 34.1 %. If we accept threshold 1/8=12.5% of the data scatter, which is the average contribution of a single entity, as the stopping criterion, then the process halts at this point. A box chart in Figure 4.3 illustrates the process. Slices corresponding to clusters are shadowed and those corresponding to splits are left blank. The most contributing features are put in split slices along with their contributions. The thinner the area of a cluster, the closer its elements to the centroid and thus to each other.

□

Example 4.35. Evolutionary tree for Gene profile data and mapping gene histories

Applying agglomerative and divisive Ward clustering, the latter with 2-Means splitting at every step, to the Gene profiles data in Table 1.3 for clustering genomes, which are columns of the table, leads to almost identical trees shown in Figure 4.4, (a) and (b), respectively; the height of each split reflects its contribution to the data scatter as described in section 5.3.2. The rows here are features. It should be noted

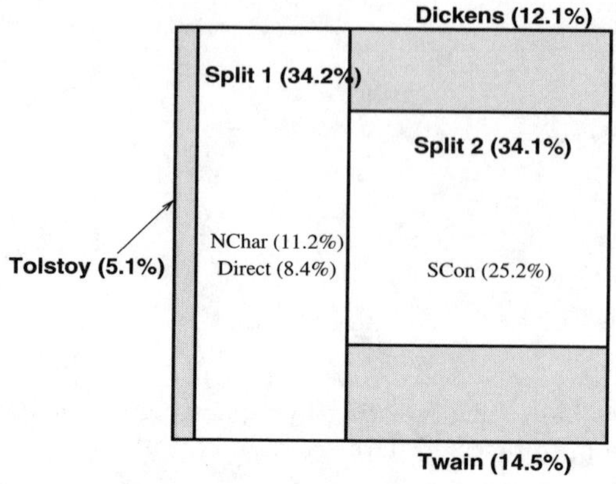

Figure 4.3: A box chart to illustrate clustering by separating; splitting was halted when split contributions to the data scatter became less than the average contribution of an individual entity.

that the first two lines in which all entries are unities do not affect the results of the computation at all, because their contributions to the data scatter are zero. Similarly, in the bacterial cluster appearing after the first split on the right, the next nine COGs (rows 3 to 11) also become redundant because they have constant values throughout this cluster.

The only difference between the two trees is the position of species b within the bacterial cluster *dcrbjqv*: b belongs to the left split part in tree (a), and to the right split part in tree (b). The first two splits after LUCA reflect the divergence of bacteria (the cluster on the right), then eukaryota (the leaf y) and archaea. All splits in these trees are compatible with the available biological knowledge; this is due to a targeted selection of COGs: of the original 1700 COGs considered in [96], more than one-third did not conform to the major divisions between bacteria, archaea and eukaryota because of extensive loss and horizontal transfer events during evolution. Due to these processes, the results obtained with a variety of tree-building algorithms on the full data are incompatible with the tree found with more robust data, such as similarities between their ribosomal proteins.

The COGs which contribute the most to the splits seem to be biologically relevant in the sense that they tend to be involved in functional systems and processes that are unique for the corresponding cluster. For instance, COG3073, COG3115, COG3107, COG2853 make the maximum contribution, of 13% each, to the split of bacteria in both trees. The respective proteins are unique to the bacterial cluster *egfs* and are bacteria-specific cell wall components or expression regulators.

Curiously, the divisive Ward-like algorithm with splitting by separating produces a different tree in which a subset of bacteria *efgsj* splits off first. In contrast to the Masterpieces set analyzed above, the procedure incorrectly determines the starting divergence here. □

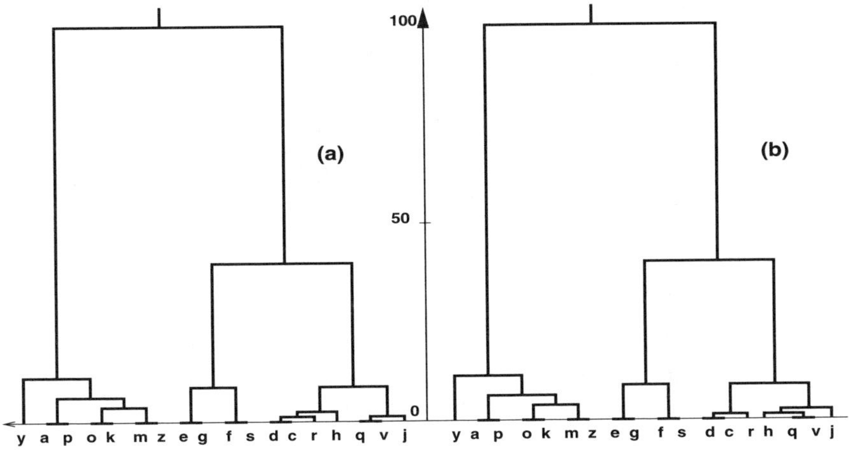

Figure 4.4: Evolutionary trees built with: (a) agglomerative Ward algorithm and (b) divisive Ward-like algorithm involving 2-Means splitting.

4.2.3 Interpretation aids for upper cluster hierarchies

Ward-like divisive clustering algorithms produce cluster trees rooted at the entity set whose leaves are not necessarily singletons, what was referred to as upper cluster hierarchies. The Ward divisive clustering model leads to a bunch of interpretation aids specifically oriented at upper cluster hierarchies [90]. No specific interpretation aids have been utilized on results of Ward agglomerative clustering. A likely reason is that agglomerative mergers have no clear classification meaning, whereas splits leading to an upper cluster hierarchy can be naturally considered as imitating building of a conceptual taxonomy. Moreover, Ward agglomerative steps contribute to the unexplained part of clustering, as will be clearly seen in Chapter 5 (see decomposition (5.28)), and thus cannot be used in the manner in which contributions to the explained part are used. In contrast, divisive steps do contribute to the explained part. Yet no interpretation aids have been conventionally utilized for the hierarchical part of an upper cluster hierarchy except for the cluster heights reflected in tree drawing.

In fact, there can be three different aspects of the contribution based interpretation aids with regard to an upper cluster hierarchy **S** because each split can be considered in terms of: (i) variables, (ii) covariances between variables, and (iii) individual entities. Let us briefly describe them in turn.

(i) **Split-to-Variable.** At this level, one can take a look at contributions of cluster splits to the total contribution of a variable to the data scatter. As stated, data scatter $T(Y) = \sum_{v \in V} T_v$ where $T_v = (y_v, y_v) = \sum_{i \in I} y_{iv}^2$ is the

Table 4.2: ScaD extended: Decomposition of the data scatter over the author-based hierarchy for Masterpieces data in Table 3.2.

Aid	Item	LenS	LenD	NChar	SCon	Pers	Obje	Dire	Total	Total,%
	Dickens	0.03	0.05	0.04	**1.17**	0.09	0.00	0.06	1.43	24.08
Typ	Twain	0.14	0.25	0.15	**0.42**	0.00	0.09	0.06	1.10	18.56
	Tolstoy	0.06	0.12	**0.50**	0.28	0.09	0.09	**0.38**	1.53	25.66
	Expl	0.23	0.41	0.69	1.88	0.18	0.18	0.50	4.06	68.30
Tax	Split1	0.09	0.16	**0.67**	0.38	0.12	0.12	**0.50**	2.03	34.22
	Split2	0.14	0.25	0.02	**1.50**	0.06	0.06	0	2.03	34.09
	Expl	0.23	0.41	0.69	1.88	0.18	0.18	0.50	4.06	68.30
	Unex	0.51	0.28	0.20	0.00	0.44	0.44	0.00	1.88	31.70
	Total	0.74	0.69	0.89	1.88	0.63	0.63	0.50	5.95	100.00

contribution of feature $v \in V$. The denotation y_v refers to column v of the pre-processed data matrix. According to (5.24) at $u = v$ one has

$$T_v = (y_v, y_v) = \sum_w \frac{N_{w1}N_{w2}}{N_w}(c_{w1,v} - c_{w2,v})^2 + (e_v, e_v) \tag{4.8}$$

where summation goes over all internal hierarchy clusters S_w split in parts S_{w1} and S_{w2}. Each split contributes, thus, $\frac{N_{w1}N_{w2}}{N_w}(c_{w1,v} - c_{w2,v})^2$; the larger a contribution the greater the variable's effect to the split.

The overall decomposition of the data scatter in (5.29),

$$T(Y) = \sum_w \frac{N_{w1}N_{w2}}{N_w} d(c_{w1}, c_{w2}) + W(S, c) \tag{4.9}$$

where S is the set of leaf clusters of an upper cluster hierarchy $\mathbf{S}$, shows contributions of both splits and leaf clusters.

Both parts of the decomposition can be used for interpretation:

– Ward distances $\frac{N_{w1}N_{w2}}{N_w} d(c_{w1}, c_{w2})$ betwen split parts, to express differences between them for taxonomic purposes, and

– decomposition of the square error criterion $W(S, c)$ over leaf clusters, which is nothing but that employed in the analysis of results of K-Means clustering in the previous Chapter; these can be used for purposes of typology rather than taxonomy.

Example 4.36. Split-to-variable aids on a box chart and in table ScaD

Interpretation split-to-variable aids are displayed at the box chart in Figure 4.3. The upper split contributes 34.2 % to the data scatter. Between cluster differences $(c_{w1v} - c_{w2v})^2$ contributing most are at variables NChar and Direct. The next split contributes 34.1% and is almost totally due to SCon (25.2 out of 34.1). A more complete picture can be seen in Table 4.2, which extends Table 3.24 ScaD by adding one more aspect: split contributions. Maximum contributions are highlighted in boldface. The upper part of the table supplies aids for typological analysis, treating each cluster as is, and the middle part for taxonomical analysis, providing aids for interpretation of splits. Both parts take into account those contributions that relate to the explained part of the leaf cluster partition. □

(ii) **Split-to-Covariance.** Feature-to-feature covariances are decomposed over splits according to cluster contributions equal to $\frac{N_{w1}N_{w2}}{N_w}(c_{w1,v} - c_{w2,v})(c_{w1,u} - c_{w2,u})$ according to (5.24), where $u, v \in V$ are two features. At this level, not only the quantities remain important for the purposes of comparison, but also one more phenomenon may occur. A covariance coefficient entry may appear with a different sign in a cluster, which indicates that in this split the association between the variables concerned changes its direction from positive to negative or vice versa. Such an observation may lead to insights into the cluster's substance.

Example 4.37. Decomposition of covariances over splits

Let us consider covariances between variables LenD, NChar, and SCon. Their total values, in thousandth, are presented in the left-hand part of the matrix equation below, and corresponding items related to the first and second splits in Figure 4.2. Entries on the main diagonal relate to the data scatter as discussed above.

	LenD	*Nch*	*Scon*			*S*1				*S*2		
LenD	87	58	−47		20	40	30		32	9	−77	
NChar	58	111	42	=	40	83	63	+	9	2	−21	+...
SCon	−47	42	234		30	63	47		−77	−21	188	

Each entry decomposed may tell us a story. For instance, the global positive correlation between NChar and SCon (+42) becomes more expressed at the first split (+63) and negative at the second split (–21). Indeed, these two are at their highest in the Tolstoy cluster and are discordant between Dickens and Twain. □

(iii) **Split-to-Entry.** Any individual row-vector y_i in the data matrix can be decomposed according to an upper cluster hierarchy **S** into the sum of items contributed by clusters $S_w \in \mathbf{S}$ containing i, plus a residual, which is zero when i itself constitutes a singleton cluster belonging to the hierarchy. This is guaranteed by the model (5.23). Each cluster S_{w1} containing i contributes the difference between its centroid and the centroid of its parent, $c_{w1} - c_w$, as described on page 153. The larger the cluster the more aggregated its contribution is.

Example 4.38. Decomposition of an individual entity over a hierarchy

The decomposition of an individual entity, such as Great Expectations (entity 3 in the Masterpieces data), into items supplied by clusters from the hierarchy presented in Figure 4.2 on page 121, is illustrated in Table 4.3. The entity data constitute the first line in this table. The other lines refer to contributions of the three largest clusters in Figure 4.2 covering entity 3: the root, the not-Tolstoy cluster, and the Dickens cluster. These clusters are formed before the splitting, after the first split, and after the second split, respectively. The last line contains residuals, due to the aggregate nature of the Dickens cluster. One can see, for instance, that SCon=0 for the entity is initially changed for the grand mean, 0.63, and then step by step degraded – the most important being the second split separating Dickens from the rest. □

Table 4.3: Single entity data decomposed over clusters containing it; the last line is the residuals.

Item	LenS	LenD	NChar	SCon	Pers	Obje	Dire
GExpectat	23.9	38	3	0	0	1	0
Grand Mean	22.45	34.13	3.00	0.63	0.38	0.38	0.25
1 split	-1.03	-3.32	-0.50	-0.13	0.13	0.13	-0.25
2 split	2.68	8.43	0.17	-0.50	0.17	-0.17	0.00
Residual	-0.20	-1.23	0.33	0.00	-0.67	0.67	0.00

This type of analysis, which emphasizes unusually high negative or positive contributions, can be applied to a wide variety of hierarchical clustering results. When an hierarchical tree obtained by clustering has substantive meaning, it also can be used for interpretation of other types of data. To illustrate this, let us consider the evolutionary trees built in the previous section and see how one can employ them to reconstruct evolutionary histories of individual genes.

Example 4.39. Using an evolutionary tree to reconstruct the history of a gene

After an evolutionary tree has been built (see Figure 4.4), one may consider the problem of finding parsimonious scenarios of gene evolution leading to the observed presence-absence patterns. An evolutionary scenario for a gene may include major evolutionary events such as emergence of the gene, its inheritance along the tree, loss, and horizontal transfer between branches of the tree. To illustrate this line of development, let us consider three COGs in Table 4.4: one from the data in Table 1.3 and two not used for building the tree, COG1514 2'-5' RNA ligase and COG0017 Aspartyl-tRNA synthetase.

Table 4.4: **Gene profiles**: Presence-absence profiles of three COGs over 18 genomes.

No	COG	Species																	
		y	a	o	m	p	k	z	q	v	d	r	b	c	e	f	g	s	j
16	COG1709	0	1	1	1	1	1	1	0	0	0	0	0	0	0	1	0	0	0
31	COG1514	0	1	0	1	1	1	1	1	1	1	0	1	0	1	1	0	0	1
32	COG0017	1	1	1	1	1	1	1	0	0	1	0	1	1	1	0	1	1	0

Given a gene presence-absence profile, let us consider, for the sake of simplicity, that each of the events of emergence, loss, and horizontal transfer of the gene is assigned the same penalty weight while events of inheritance along the tree are not penalized as conforming to the tree structure. Then, COG1709 (the first line of Table 4.4) can be thought of as having emerged in the ancestor of all archaea (node corresponding to cluster *aompkz* and then horizontally transferred to species *f*, which amounts to two penalized events. No other scenario for this COG has the same or smaller number of events.

The reconstructed history of COG1514 (the second line of Table 4.4) is more complicated. Let us reconstruct it on the right-hand tree of Figure 4.4. Since this COG is represented in all non-singleton clusters, it might have emerged in the root of the tree. Then, to conform to its profile, it must be lost at y, o, s, g, c and r, which amounts to 7 penalized events, one emergence and six losses, altogether. However, there exist better scenarios with six penalized events each. One such scenario assumes that the gene emerged in the node of cluster archaea *aompkz* and then horizontally transferred to cluster *vqbj* and singletons f, e, d, thus leading to one emergence event, four horizontal transfers, and one loss, at o. Obviously, the profile data by themselves give no clue to the node of emergence. In principle, the gene in question might have emerged in cluster *vqbj* or even in one of the singletons. To resolve this uncertainty, one needs additional data on evolution of the given gene, but this is beyond the scope of this book. □

4.3 Conceptual clustering

This method of divisive clustering builds an upper cluster hierarchy by sequentially splitting clusters, each time using a single attribute rather than all of them as in the previous section. Such is the hierarchy of Digits in Figure 1.5 built by splitting the entire set according to the presence or absence of the bottom edge $e7$ and then splitting Yes entities according to the presence or absence of $e5$ and $e1$.

Originally, the method was developed for categorical features only. A goodness-of-split criterion is utilized to decide what class S in a hierarchy to split and by what variable. To define such a criterion, various approaches can be used. Let us consider two popular approaches that are compatible with the data recovery framework.

1. **Impurity.** Denote a set of entities by $J \subseteq I$ and a qualitative variable on it by l, with p_v denoting frequencies of its categories $v \in V_l$ in J. Let us measure the dispersion of l on J with the *Gini coefficient*, $G(l) = 1 - \sum_{v \in V_l} p_v^2 = \sum_{v \in V_l} p_v(1 - p_v)$ defined in section 2.1.3. If J is divided in clusters $\{S_k\}$, this cross classifies l so that any category $v \in V_l$ may co-occur with any S_k; let us denote the frequency of the co-occurrence by p_{kv}. Then the Gini coefficient for l over S_k will be $G(l/k) = 1 - \sum_{v \in V_l} p(v/k)^2$ where $p(v/k) = p_{kv}/p_k$ with p_k denoting proportion of entities of J in S_k. The average change of the Gini coefficient after the split of J into S_k, $k = 1, ..., K$, can be expressed as the difference, $\Delta(l, S) = G(l) - \sum_k p_k G(l/k)$. It is this expression which is referred to as the impurity function in [11]. It also equals the summary Quetelet index G^2 according to the analysis on page 56. The greater the $\Delta(l, S)$ the better the split S.

2. **Category utility.** Consider a partition $S = \{S_k\}$ $(k = 1, ..., K)$ on J and a set of categorical features $l \in L$ with categories $v \in V_l$. The category utility function scores partition S against the set of variables according to formula [31]:

$$u(S) = \frac{1}{K}\sum_{k=1}^{K} p_k[\sum_l \sum_{v\in V_l} P(l=v|S_k)^2 - \sum_l \sum_{v\in V_l} P(l=v)^2] \qquad (4.10)$$

The term in square brackets is the increase in the expected number of attribute values that can be predicted given a class, S_k, over the expected number of attribute values that could be predicted without using the class. The assumed prediction strategy follows a probability-matching approach. According to this approach, category v is predicted with the frequency reflecting its probability, $p(v/k)$ within S_k, and $p_k = N_k/N$ when information of the class is not provided. Factors p_k weigh classes S_k according to their sizes, and the division by K takes into account the difference in partition sizes: the smaller the better.

Either of these functions can be applied to a partition S to decide which of its clusters is to be split and how. They are closely related to each other as well as to contributions B_{kv} in section 3.4.2, which can be stated as follows.

Statement 4.4. *The impurity function $\Delta(l, S)$ equals the summary contribution $B(l, S) = \sum_{v\in V_l}\sum_{k=1}^{K} B_{kv}$ with scaling factors $b_v = 1$ for all $v \in V_l$. The category utility function $u(S)$ is the sum of impurity functions over all features $l \in L$ related to the number of clusters K, $u(S) = \sum_l \Delta(l, S)/K$.*

Proof: Indeed, according to the definition of impurity function, $\Delta(l, S) = G(l) - \sum_k p_k G(l/k) = 1 - \sum_{v\in V_l} p_v^2 - \sum_k p_k(1 - \sum_{v\in V_l} p(v/k)^2) = \sum_k \sum_v \frac{p_{kv}}{p_k} - \sum_{v\in V_l} p_v^2 = B(l, S)$. To prove the second part of the statement, let us note that $P(l = v|S_k) = p(v/k)$ and $P(l = v) = p_v$. This obviously implies that $u(S) = \sum_l \Delta(l, S)/K$, which proves the statement.

The summary impurity function, or the category utility function multiplied by K, is exactly the summary contribution of variables l to the explained part of the data scatter, $B(S, c)$, that is, the complement of the K-Means square error clustering criterion to the data scatter, when the data pre-processing has been done with all $b_v = 1$ ($v \in V_l$) [93]. In brief, maximizing the category utility function is equivalent to minimizing the K-Means square error criterion divided by the number of clusters with the data standardized as described above.

This invites, first, different splitting criteria associated with $B(S, c)$ at different rescaling factors b_v and b'_v, and, second, extending the splitting criterion to the mixed scale feature case, which is taken into account in the following formulation.

Conceptual clustering with binary splits
1. *Initial setting.* Set S to consist of the only cluster, the entire entity set I.
2. *Evaluation.* In the cycle over all clusters k and variables l, consider the possibility of splitting S_k over l in two parts. If l is quantitative, the split must correspond to a split of its range in two intervals: if bl and br are the minimum and maximum values of l at S_k, take a number T of locations for the splitting point, $p_t = bl + t(br - bl)/T$ $(t = 1, ..., T)$, and select that one of them which maximizes the goodness-of-split function. If l is nominal and has more than two categories, consider categories $v \in V_l$ as entities weighted by their frequencies in S_k, $p(v/k)$, and apply a version of the correspondingly modified Serial splitting algorithm to divide them in two parts.
3. *Splitting.* Select the pair (k, v) that received the highest score and do the binary split.
4. *Halt.* Check stop-condition. If it is satisfied, end. Otherwise go to 2.

Since the scoring function is the same as in the Ward-like divisive clustering of the previous section, the stop-condition here can be borrowed from that on page 117. One more stopping criterion comes from the category utility function, which is the total contribution divided by the number of clusters K: calculations should stop when this goes down. Unfortunately, this simple idea seems to not always work, as will be seen in the following examples.

Example 4.40. Conceptual clustering of Digits

The classification tree of the Digit data in Figure 1.5 on page 16 has been produced with the process above, assuming all the features are binary nominal. Let us take, for instance, partition $S = \{S_1, S_2\}$ of I according to attribute $e2$ which is present at S_1 comprising 4,5, 6, 8, 9, and 0, and is absent at S_2 comprising 1,2,3, and 7. Cross-classification of S and $e7$ in Table 4.5 yields $\Delta(e7, S) = 0.053$.

Table 4.5: Cross-tabulation of S (or, e2) against e7.

e7	S_1	S_2	Total
e7=1	5	2	7
e7=0	1	2	3
Total	6	4	10

To see what this has to do with the setting in which K-Means complementary criterion applies, let us pre-process the Digit data matrix by subtracting the averages within each column (see Table 4.6); note that the scaling coefficients are all unity here.

However, the data in Table 4.6 is not exactly the data matrix Y considered theoretically because both Y and X must have 14 columns after enveloping each of the

Table 4.6: Data in Table 2.3 1/0 coded with the follow-up centering of the columns.

e1	e2	e3	e4	e5	e6	e7
-.8	-.6	.2	-.7	-.4	.1	-.7
.2	-.6	.2	.3	.6	-.9	.3
.2	-.6	.2	.3	-.4	.1	.3
-.8	.4	.2	.3	-.4	.1	-.7
.2	.4	-.8	.3	-.4	.1	.3
.2	.4	-.8	.3	.6	.1	.3
.2	-.6	.2	-.7	-.4	.1	-.7
.2	.4	.2	.3	.6	.1	.3
.2	.4	.2	.3	-.4	.1	.3
.2	.4	.2	-.7	.6	.1	.3

14 categories reflected in Table 2.3. Columns corresponding to the category "ei is absent" in all features i=1,2,...,7 are not included in Table 4.6, because they provide no additional information.

The data scatter of this matrix is the summary column variance times $N = 10$, which is 13.1. However, to get the data scatter in the lefthand side of (5.13), this must be doubled to 26.2 to reflect the "missing half" of the virtual data matrix Y.

Let us now calculate within class averages c_{kv} of each of the variables, $k = 1, 2$, v=e1,...,e7, and take contributions $N_k c_{kv}^2$ summed up over clusters S_1 and S_2. This is done in Table 4.7, the last line in which contains contributions of all features to the explained part of the data scatter.

Table 4.7: Feature contributions to digit classes defined by e2.

e2	e1	e2	e3	e4	e5	e6	e7
e2=1	0.007	0.960	0.107	0.107	0.060	0.060	0.107
e2=0	0.010	1.440	0.160	0.160	0.090	0.090	0.160
Total	0.017	2.400	0.267	0.267	0.150	0.150	0.267

The last item, 0.267, is the contribution of e7. Has it anything to do with the reported value of impurity function $\Delta(e7, S) = 0.053$? Yes, it does. There are two factors that make these two quantities different. First, to get the contribution from Δ it must be multiplied by $N = 10$ leading to $10\Delta(e7, S) = 0.533$. Second, this is the contribution to the data scatter of matrix Y obtained after enveloping all 14 categories which has not been done in Table 4.6, thus, not taken into account in the contribution 0.267. After the contribution is properly doubled, the quantities do coincide.

Similar calculations made for the other six attributes, e1, e2, e3,..., e6, lead to the total $\sum_{l=1}^{7} \Delta(el, S) = 0.703$ and, thus, to $u(S) = 0.352$ according to Statement 4.4. since $M = 2$. The part of the data scatter taken into account by partition S is the total of $\Delta(S, el)$ over $l = 1, ..., 7$ times $N = 10$, according to (5.22), that is, 7.03 or 26.8% of the scatter 26.2.

The evaluations at the first splitting step of the total Digit set actually involve all pairwise contingency coefficients $G^2(l, l') = \Delta(l, l')$ $(l, l' = 1, ..., 7)$ displayed in Table 4.8. According to this data, the maximum summary contribution is supplied by the S made according to e7; it is equal to 9.63 which is 36.8% of the total data scatter.

Table 4.8: Pairwise contingency coefficients. In each column el values of Δ for all variables are given under the assumption that partition S is made according to el (l=1,...,7).

Target	e1	e2	e3	e4	e5	e6	e7
e1	0.320	0.003	0.020	0.015	0.053	0.009	0.187
e2	0.005	0.480	0.080	0.061	0.030	0.080	0.061
e3	0.020	0.053	0.320	0.034	0.003	0.009	0.034
e4	0.020	0.053	0.045	0.420	0.003	0.020	0.115
e5	0.080	0.030	0.005	0.004	0.480	0.080	0.137
e6	0.005	0.030	0.005	0.009	0.030	0.180	0.009
e7	0.245	0.053	0.045	0.115	0.120	0.020	0.420
Total	0.695	0.703	0.520	0.658	0.720	0.398	0.963

Thus, the first split must be done according to e7. The second split, according to e5, contributes 3.90, and the third split, according to el, 3.33, so that the resulting four-class partition, $S = \{\{1, 4, 7\}, \{3, 5, 9\}, \{6, 8, 0\}, \{2\}\}$, contributes 9.63+3.90+3.33=16.87 =64.4% to the total data scatter. The next partition step would contribute less than 10% of the data scatter, that is, less than an average entity, which may be considered a signal to stop the splitting process.

One should note that the category utility function $u(S)$ after the first split is equal to 9.63/2=4.81, and after the second split, to (9.63+3.90)/3=13.53/3=4.51. The decrease means that calculations must be stopped after the very first split, according to the category utility function, which is not an action of our preference. □

Example 4.41. Relation between conceptual and Ward clustering of Gene profiles

The divisive tree of species according to gene profiles on Figure 4.4 (b) can be used for analysis of the conceptual clustering category utility score criterion (4.10) which is equivalent to the ratio of the explained part of the data scatter over the number of clusters. Indeed, the first split contributes 50.3% to the data scatter, which makes the category utility function u be equal to $50.3/2 = 25.15$. The next split, of the bacterial cluster, adds 21.2%, making the total contribution 50.3+21.2=71.5%, which decreases the utility function to $u = 71.5/3 = 23.83$. This would force the division process to stop at just two clusters, which shows that the normalizing value K might be overly stringent.

This consideration may be applied not only to the general divisive clustering results in Figure 4.4 (b) but to conceptual clustering results as well. Why? Because each of the two splits, although found with the multidimensional search, also can be done monothetically, with one feature only: the first split at COG0290 or COG1405 (lines 4, 17 in Table 1.3) and the second one at COG3073 or COG3107 (lines 22, 28 Table 1.3). These splits must be optimal because they have been selected in the much less restrictive multidimensional splitting process. □

Among other goodness-of-split criteria considered for categorical variables, the chi-squared is utilized in CHAID [42], the entropy in C4.5 [111], and the so-called twoing rule in CART [11].

4.4 Extensions of Ward clustering

4.4.1 Agglomerative clustering with dissimilarity data

Given a dissimilarity matrix $D = (d_{ij})$, $i, j \in I$, the Ward algorithm can be reformulated as follows.

Ward dissimilarity agglomeration:
1. *Initial setting.* All the entities are considered as singleton clusters so that the between-entity dissimilarities are between-cluster dissimilarity. All singleton heights are set to be zero.
2. *Agglomeration rule.* Two candidate clusters, S_{w1} and S_{w2}, that are nearest to each other (that is, being at the minimum distance) are merged together forming their parent cluster $S_{w1\cup w2} = S_{w1} \cup S_{w2}$, and the merged cluster's height is defined as the sum of the children's heights plus the distance between them.
3. *Distance.* If the newly formed cluster $S_{w1\cup w2}$ coincides with the entire entity set, go to Step 4. Otherwise, remove S_{w1} and S_{w1} from the set of candidate clusters and define distances between the new cluster $S_{w1\cup w2}$ and other clusters S_k as follows:

$$dw_{w1\cup w2,k} = \frac{N_{w1}+N_k}{N^+}dw_{w1,k} + \frac{N_{w2}+N_k}{N^+}dw_{w2,k} - \frac{N_k}{N^+}dw_{w1,w2} \quad (4.11)$$

where $N^+ = N_{w1\cup w2} + N_k$. The other distances remain unvaried. Then, having the number of candidate clusters reduced by one, go to Step 2.
4. *Output.* Output upper part of the cluster tree according to the height function.

It can be proven that distance (4.11) is equal to Ward distance between the merged cluster and other clusters when D is a matrix of Euclidean squared distances. In fact, formula (4.11) allows the calculation and update of Ward distances without calculation of cluster centroids.

Agglomeration step 2 remains computationally intensive. However, the amount of calculations can be decreased because of properties of the Ward distance [103].

4.4.2 Hierarchical clustering for contingency and flow data

The contingency data format has been introduced in section 2.2.3. Due to the fact that contingency data are not only measured in the same scale but also measure different parts of the data flow and thus can be summed up to the

Table 4.9: Confusion contingency data.

Entity	Feature 1	2	3	4	5	6	7	8	9	0	Total
1	0.088	0.001	0.001	0.002	0.000	0.002	0.006	0	0.000	0.000	0.100
2	0.001	0.078	0.005	0.000	0.004	0.005	0.001	0.003	0.001	0.002	0.100
3	0.003	0.003	0.068	0.100	0.002	0.000	0.004	0.003	0.015	0.002	0.100
4	0.015	0.002	0.000	0.073	0.000	0.001	0.003	0.001	0.004	0	0.100
5	0.001	0.003	0.004	0.001	0.067	0.008	0.001	0.001	0.013	0.001	0.100
6	0.002	0.001	0.001	0.001	0.010	0.063	0.000	0.016	0.001	0.004	0.100
7	0.027	0.000	0.002	0.002	0.001	0	0.067	0	0.000	0.001	0.100
8	0.001	0.003	0.003	0.002	0.002	0.007	0.001	0.058	0.007	0.017	0.100
9	0.002	0.003	0.011	0.005	0.008	0.001	0.002	0.008	0.055	0.004	0.100
0	0.002	0.000	0.001	0.001	0.001	0.002	0.002	0.007	0.002	0.082	0.100
Total	0.143	0.095	0.096	0.089	0.094	0.088	0.088	0.096	0.098	0.113	1.000

total number of observations, they can be processed with a greater extent of comparability than the ordinary entity-to-feature data.

To introduce the concepts needed, let us consider a contingency table $P = (p_{tu})$, $t \in T$, $u \in U$, whose entries have been divided by the total flow p_{++}, which means that $p_{++} = 1$. The marginals p_{t+} and p_{+u}, which are just within-row and within-column totals, will be referred to as the weights of rows $t \in T$ and columns $u \in U$.

For any subset of rows $S \subseteq T$, the conditional probability of a column u can be defined as $p(u/S) = p_{Su}/p_{S+}$ where p_{Su} is the sum of frequencies p_{tu} over all $t \in S$ and p_{S+} the summary frequency of rows $t \in S$.

Example 4.42. Aggregating Confusion data

For the Confusion data in Table 1.9, the matrix of relative frequencies is in Table 4.9. For example, for $S = \{1, 4, 7\}$ and $u = 3$, $p_{S3} = 0.001 + 0.000 + 0.002 = 0.003$ and $p_{S+} = 0.100 + 0.100 + 0.100 = 0.300$ so that $p(3/S) = 0.003/0.300 = 0.010$. Analogously, for $u = 1$, $p_{S1} = 0.088 + 0.015 + 0.027 = 0.130$ and $p(1/S) = 0.130/0.100 = 1.30$. □

The row set S will be characterized by its profile, the vector of conditional probabilities $g(S) = (p(u/S))$, $u \in U$.

Then, the chi-squared distance between any two non-overlapping row sets, S_1 and S_2, is defined as

$$\chi(g(S_1), g(S_2)) = \sum_{u \in U} (p(u/S_1) - p(u/S_2))^2 / p_{+u} \tag{4.12}$$

Using this concept, Ward's agglomeration algorithm applies to contingency data exactly as it has been defined in section 4.1 except that the Ward distance is modified here to adapt to the situation when both rows and columns are weighted:

$$w(S_{h1}, S_{h2}) = \frac{p_{S_{h1}} + p_{S_{h2}}}{p_{S_{h1} \cup S_{h2}}} \chi(g(S_{h1}), g(S_{h2})) \tag{4.13}$$

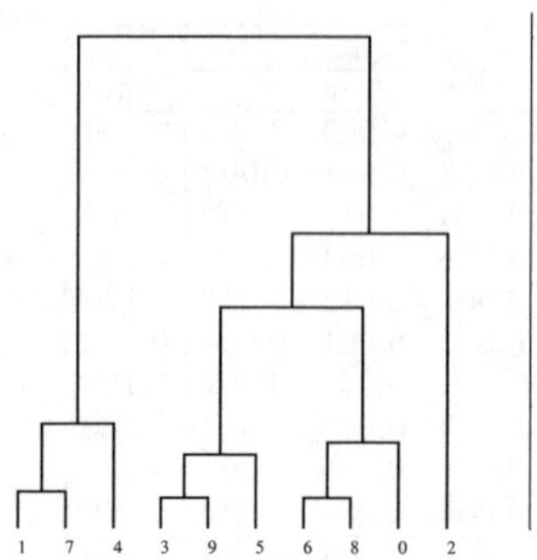

Figure 4.5: The hierarchy found with the modified Ward algorithm for Confusion data.

This definition differs from the standard definition in the following three aspects:

1. Profiles are taken as cluster centroids.

2. Chi-squared distance is taken instead of the Euclidean squared.

3. Marginal frequencies are used instead of cardinalities.

These formulas are derived in section 5.4.3 from a data recovery model relating Quetelet coefficients in the original data table and that aggregated according to the clustering. It appears, they express the decrement of the Pearson chi-square contingency coefficient under the aggregation.

Example 4.43. Clustering and aggregation of Confusion data

The drawing in Figure 4.5 shows the hierarchy of Digits row clusters found with the agglomerative clustering algorithm which uses the chi-square distance (4.12). Curiously, the same topology, with slightly changed heights, emerges when the data table is aggregated over rows and columns simultaneously to minimize the decrement of $X^2(F,F)$ of the aggregated table.

The aggregate confusion rates and Quetelet coefficient data corresponding to the four-class partition, $S = \{\{1,4,7\},\{3,5,9\},\{6,8,0\},\{2\}\}$, are on the right in Table 4.10. □

Table 4.10: **Confusion:** Four cluster aggregate Confusion data and corresponding Quetelet coefficients.

I	2827	33	96	44	I	1.95	-0.88	-0.89	-0.95
II	33	783	90	94	II	-0.90	7.29	-0.69	-0.68
III	203	85	2432	280	III	-0.79	-0.70	1.81	-0.69
IV	134	46	263	2557	IV	-0.84	-0.84	-0.70	1.86

4.5 Overall assessment

Advantages of hierarchical clustering:

1. Visualizes the structure of similarities in a convenient form.
2. Models taxonomic classifications.
3. Provides a bunch of interpretation aids at the level of entities, variables and variable covariances.

Less attractive features of the approach:

1. Massive computations related to finding minimum distances at each step, which is especially time consuming in agglomerative algorithms.
2. Rigidity: Having a splitting or merging step, no possibility to change it afterwards.

Computations in agglomerative clustering can be drastically reduced if the minima from previous computations are kept.

Chapter 5

Data Recovery Models

Main subjects covered:

1. What is the data recovery approach.
2. A data recovery model and method for Principal component analysis.
3. A data recovery model and data scatter decomposition for K-Means and Anomalous cluster clustering.
4. A data recovery model and data scatter decompositions for cluster hierarchies.
5. A unified matrix equation model for all three above.
6. Mathematical properties of the models justifying methods presented in previous chapters.
7. Extensions of the models, criteria and methods to similarity and contingency data.
8. One-by-one data recovery clustering methods.
9. Data recovery interpretation of correlation and association coefficients.

Base words

Alternating optimization A method in the theory of optimization, applicable when a function to be optimized depends on two or more groups of variables. The method works iteratively by optimizing the function over a group of variables having the other group specified. K-Means is a method of alternating minimization of the square error clustering criterion.

Anomalous cluster A cluster which is furthest from the reference point. The iterated Anomalous clustering implements the one-by-one separating strategy of Principal Component Analysis in the data recovery model for K-Means, and thus allows for extending K-Means to its intelligent version, iK-Means, mitigating the need in defining an initial setting.

Attraction coefficient A measure of attraction of an entity to a cluster, which is equal to the entity's average similarity to the cluster minus half of the average within cluster similarity. In K-Means and Ward clustering, attractions of entities to their clusters are greater than to other clusters. Moreover, in K-Means, entities are always positively attracted to their clusters and negatively to other clusters.

Contingency data A data table whose rows and columns correspond to two sets of categories and entries are counts or proportions of observations at the intersection of a row and column categories (co-occurrence values). Because of their summability across the table, contingency data are treated after having been transformed into relative Quetelet coefficients. The data scatter of the transformed data is measured by the chi-square contingency coefficient.

Correspondence factor analysis A PCA-like method for visualization of co-occurrence values in a contingency table by displaying both row and column items as points in the same 2D or 3D space. Correspondence factor analysis is a data recovery method in which the recovered entries are Quetelet association coefficients. The method heavily relies on the singular value decomposition of a related matrix.

Data recovery models in clustering A data recovery model includes a rule which gives a value, according to a cluster structure, to every data entry. This way, every datum is represented as the sum of a cluster-recovered value and a residual, which provides for a built-in quality principle: the smaller the residuals, the better the cluster structure.

Data scatter decomposition A decomposition of the data scatter in two parts, that explained by the cluster structure and that remaining unexplained. Such a decomposition provides for both an explicit cluster criterion and interpretation aids, the former being the minimum of the unexplained part and the latter various parts of the explained part. The explained part in K-Means and Ward clustering is always the sum of contributions of individual clusters or splits.

Linear regression A method for analysis of interrelation between two quantitative features x and y in which y is approximated by an affine transformation $ax + b$ of x, where a and b are referred to as slope and intercept, respectively. This setting is the genuine ground on which the coefficients of correlation and determination are defined and substantiated.

One-by-one clustering A method in clustering in which clusters or splits are taken one by one. In this text, all such methods exploit the additive structure of clustering data recovery models, which is analogous to that of the model of Principal Component Analysis. Cluster separations and splits are to be made in the order of their contribution to the data scatter. In this way, the additive structure of the data scatter decomposition is maintained to provide for model-based interpretation aids.

Principal component analysis A method for approximation of a data matrix with a small number of hidden factors, referred to as principal components, such that data entries are expressed as linear combinations of hidden factor scores. It appears that principal components can be determined with the singular value decomposition (SVD) of the data matrix.

Reference point A vector in the variable space serving as the space origin. The Anomalous pattern is sought starting from an entity furthest from the reference point, which thus models the norm from which the Anomalous pattern deviates most.

Split versus separation Difference between two perspectives: cluster-versus-the-rest and cluster-versus-the-whole, reflected in different coefficients attached to the distance between centroids of split parts. The former perspective is taken into account in the Ward-like divisive clustering methods, the latter in the Anomalous pattern clustering.

Ward-like divisive clustering A divisive clustering method using Ward distance as the splitting criterion. The method can be considered an implementation of the one-by-one PCA strategy within the data recovery clustering.

5.1 Statistics modeling as data recovery

The data recovery approach is the cornerstone of contemporary thinking in statistics and data analysis. It is based on the assumption that the observed data reflect a regular structure in the phenomenon of which they inform. The regular structure A, if known, would produce data F(A) that should coincide with the observed data Y up to small residuals which are due to possible flaws in any or all of the following three aspects: (a) sampling entities, (b) selecting features and tools for their measurements, and (c) modeling the phenomenon in question. Each of these can drastically affect results. However, so far only the simplest of the aspects, (a), has been addressed scientifically by introduction of probabilities to study the significance of statistical inference in data analysis. In this treatise we are not concerned with these issues. We are concerned with the underlying equation:

$$\textbf{Observed data Y} = \textbf{Recovered data F(A)} + \textbf{Residuals E} \qquad (*)$$

The quality of the model A is assessed according to the level of residuals E: the smaller the residuals the better the model. Since quantitative models involve unknown coefficients and parameters, this naturally leads to the idea of fitting these parameters to data in such a way that the residuals become as small as possible. To put this idea as a minimization problem, one needs to combine the multiple residuals in an aggregate criterion. In particular, the so-called principle of maximum likelihood has been developed in statistics. When the data can be modelled as a random sample from a multivariate Gaussian distribution, this principle leads to the so-called least squares criterion, the sum of squared residuals to be minimized. In the data mining framework, the data do not necessarily come from a probabilistic population. Moreover, analysis of the mechanism of data generation is not of primary concern. One needs only to see if there are any patterns in the data as they are. In this case, the principle of maximum likelihood may be not applicable. Still, the sum of squared residuals criterion can be used in the context of data mining as a clear cut measure of the largeness of the residuals. It provides for nice geometric properties and guaranties the production of provably reasonable cluster solutions. It admits useful decompositions of the data scatter into the sum of explained and unexplained parts. No assumptions of particular distributions for the residuals are exploited in the further treatment. To show the working of model (∗) along with the least-squares principle, let us introduce four examples covering important methods in data mining: (a) averaging, (b) linear regression, (c) principal component analysis, and (d) correspondence analysis. These examples are also used to introduce some useful concepts in data analysis that are used throughout this text.

5.1.1 Averaging

Let a series of real numbers, $x_1, ..., x_N$, have been assumed to represent the same unknown value a. Equation (*) then becomes

$$x_i = a + e_i$$

with e_i being the residual for $i = 1, ..., N$. To minimize the sum of squares $L(a) = \sum_i e_i^2 = \sum_i (x_i - a)^2$ as a function of a, one may utilize the first-order optimality condition, $dL/da = 0$, that is, $dL/da = -2\sum_i x_I - Na = 0$. That means that the least-squares solution is the average $a = \bar{x} = \sum_i x_i/N$. By substituting this for a in $L(a)$, one obtains $L(\bar{x}) = \sum_i x_i^2 - N\bar{x}^2$. The last expression gives, in fact, the decomposition of the data scatter, the sum of data entries squared $T(x) = \sum_i x_i^2$, into the explained and unexplained parts, $T(x) = N\bar{x} + L(\bar{x})$. The averaged unexplained value $L(\bar{x})/N$ is the well known variance $s(x)^2$ of the series, and its square root, $s(x) = \sqrt{L(\bar{x})/N}$, the standard deviation. It appears thus that the average minimizes the standard deviation $s(x)$ of observations from a.

5.1.2 Linear regression

Let a series of pairs of reals, $(x_1, y_1), ..., (x_N, y_N)$ such as IQ score for x_i and math mark for y_i at individual i ($i = 1, ..., N$) have been collected. The linear regression model assumes that y-values are effected by x-values according to a linear equation $y = ax + b$ where a and b are constant coefficients, referred to as the slope and intercept, respectively. To fit the values of a and b to data, the traditional thinking considers that only y-values are to be explained by model (*), thus leading to equations

$$y_i = ax_i + b + e_i.$$

The least squares criterion for minimizing the residuals in this case is a function of two unknown coefficients, $L = L(a, b) = \sum_i e_i^2 = \sum_i (y_I - ax_i - b)^2$. The first-order optimality conditions lead to $a = \sum_i (x_i - \bar{x})(y_i - \bar{y})/s^2(x)$ and $b = \bar{y} - a\bar{x}$. The linear regression $y = ax + b$ with a and b fitted to the data can be used for analysis and prediction of y, given x, if L is small.

A symmetric function of features x and y, the correlation coefficient, has been defined as $\rho = \sum_i (x_i - \bar{x})(y_i - \bar{y})/[Ns(x)s(y)]$. The optimal a thus can be expressed through ρ as $a = \rho s(y)/s(x)$. By putting the optimal a and b into $L(a, b)$, the minimum L can be expressed as $L = Ns^2(y)(1 - \rho^2)$.

According to the formulas above, the correlation coefficient ρ in the data recovery paradigm has the following properties:

1. Its square, the so-called determination coefficient ρ^2, expresses the decrease of the variance of y after its linear relation to x has been taken into account.

2. The values of ρ are restricted to the interval between -1 and 1. The closer ρ is to either 1 or -1, the smaller are the residuals in equation $(*)$. For instance, at $\rho = 0.9$, the unexplained variance of y constitutes $1 - \rho^2 = 19\%$ of its original variance.

3. The slope a is proportional to ρ so that a is positive or negative depending on the sign of correlation coefficient. When $\rho = 0$ the slope is 0 too; the variables y and x are referred to as non-correlated, in this case, which means that there is no linear relation between them, though another functional relation, such as a quadratic one, may exist. The case of $\rho = 0$ geometrically means that centered versions of feature vectors $x = (x_i)$ and $y = (y_i)$ are mutually orthogonal.

4. With the data pre-processed as

$$x' = \frac{x - \bar{x}}{s(x)} \text{ and } y' = \frac{y - \bar{y}}{s(y)} \tag{5.1}$$

the variances become unities, thus leading to simpler formulas: $a' = \rho = (x', y')/N$, $b' = 0$, $L' = N(1 - \rho^2)$, and $N = N\rho^2 + L'$ where N happens to be equal to the scatter of y'.

5. The value of ρ does not change under linear transformations of scales of x and y.

5.1.3 Principal component analysis

Principal component analysis[1] is a major tool for approximating observed data with model data formed by a few 'hidden' factors. Observed data such as marks of students $i \in I$ at subjects labelled by $l = 1, ..., M$ constitute a data matrix $X = (x_{il})$. Assume that each mark x_{il} reflects the student's hidden ability z_i $(i \in I)$ with an impact coefficient c_l, due to subject l's specifics. The principal component analysis model suggests that the student i's score over subject l reflects the product of the mutual impact of student and subject, $z_i c_l$. Then equation $(*)$ can be formulated as

$$x_{il} = c_l z_i + e_{il}. \tag{5.2}$$

The least squares criterion is $L = \sum_{i \in I} \sum_{l \in L} (x_{il} - c_l z_i)^2$ and the first-order optimality conditions lead to equations $\sum_l x_{il} c_l = z_i \sum_l c_l^2$ and $\sum_i x_{il} z_i = c_l \sum_i z_i^2$ for all $l \in L$ and $i \in I$. Sums $\sum_l c_l^2$ and $\sum_i z_i^2$ are squared norms of vectors c and z with the norms being defined as $||c|| = \sqrt{\sum_l c_l^2}$ and $||z|| =$

[1] This section, as well as the next one, can be understood in full only if introductory concepts of linear algebra are known, including the concepts of matrix and its rank.

$\sqrt{\sum_i z_i^2}$. A vector whose norm is unity is referred to as a normed vector; to make $z = (z_i)$ normed, z has to be divided by its norm: vector $z' = z/||z||$ is the normed version of z. Let us denote by μ the product $\mu = ||z||||c||$ and by z^* and c^* the normed versions of the least-squares solution c, z. Then the equations above can be rewritten as $\sum_l x_{il} c_l^* = \mu z_i^*$ and $\sum_i x_{il} z_i^* = \mu c_l^*$, or in matrix algebra notation, $Xc^* = \mu z^*$ and $X^T z^* = \mu c^*$. These are quite remarkable equations expressing the fact that optimal vectors c and z are linear combinations of, in respect, rows and columns of matrix X. These expressions define an inherent property of matrix X, its singular value and vectors, where μ is a singular value of matrix X and c^* and z^* are the normed singular vectors corresponding to μ. It is well known that the number of non-zero singular values of a matrix is equal to its rank and, moreover, the singular vectors c^* corresponding to different singular values are mutually orthogonal, as well as the vectors z^* [38]. In our case however, μ must be the maximum singular value of X because of the decomposition of the data scatter $T(X) = \sum_{i,l} x_{il}^2$,

$$T(X) = \mu^2 + L, \tag{5.3}$$

that holds for the optimal c^* and z^*. Indeed, since the data scatter $T(X)$ is constant if the data do not change, the unexplained part L is minimum when the explained part, μ^2, is maximum. To derive decomposition (5.3), one needs to perform the squaring operation in the formula for L and take into account the fact that $\sum_{i,l} x_{il} z_i c_l = ||c||^2||z||^2$ for the optimal c^* and z^*, which can be proven by multiplying each i-th equation $\sum_l x_{il} c_l = z_i \sum_l c_l^2$ by z_i and summing up the results over $i \in I$. The talent score vector z^* is referred to as the principal component and the corresponding c^* as the loading vector.

To keep up with the model (5.2), vectors z^* and c^* must be rescaled to contain μ in their product, which is usually done with formulas $c = \sqrt{\mu} c^*$ and $z = \sqrt{\mu} z^*$.

Generalizing the one-factor model (5.2), one may assume a small number m of different hidden "talent" factors z_1,..., z_m forming an unknown $N \times m$ score matrix Z_m with corresponding m loading vectors c_1,..., c_m forming rows of an $m \times M$ loading matrix C_m so that equation (5.2) is extended to

$$x_{il} = c_{1l} z_{i1} + ... + c_{ml} z_{im} + e_{il} \tag{5.4}$$

for all $i \in I$ and $l = 1, ..., M$, or, in the matrix algebra notation,

$$X = Z_m C_m + E \tag{5.5}$$

We are interested in finding the least squares solution to equation (5.4) – (5.5), that is, matrices Z_m and C_m minimizing the sum of squared elements of residual matrix E. It is not difficult to see that the solution is not unique. Indeed, any solution Z_m, C_m can be transformed to $Z = Z_m F^T$ and $C = FC_m$ with any

$m \times m$ matrix F satisfying equation $F^T = F^{-1}$, that is, being a "rotation" matrix [27, 38]. Obviously $ZC = Z_m C_m$, that is, the rotated solution Z, C corresponds to the same residual matrix E and thus the same value of L. This shows that the least-squares solution is defined only up to the linear subspace of the space of N-dimensional vectors, whose base is formed by columns of matrix Z_m.

The optimal linear subspace can be specified in terms of the so-called singular value decomposition (SVD) of matrix X. Let us recall that SVD of $N \times M$ matrix X amounts to equation $X = Z\Lambda C$ where Z is $N \times r$ matrix of mutually orthogonal normed N-dimensional column-vectors z_k^*, C is $r \times M$ matrix of mutually orthogonal normed M-dimensional row-vectors c_k^*, and Λ is diagonal $r \times r$ matrix with positive singular values μ_k on the main diagonal such that z_k^* and c_k^* are normed singular vectors of X corresponding to its singular value μ_k so that $Xc_k^* = \mu_k z_k^*$ and $X^T z_k^* = \mu_k c_k^*$, $k = 1, ..., r$ [38]. The SVD decomposition is proven to be unique when singular values μ_t are mutually different. A least-squares solution to model (5.5) can now be determined from matrices Z_m^* and C_m^* of m singular vectors z_k^* and c_k^* corresponding to m greatest singular values μ_k, $k = 1, ..., m$. (The indices reflect the assumption that the singular values have been placed in the order of descent, $\mu_1 \geq \mu_2 \geq ... \geq \mu_r > 0$.) Let us denote the diagonal matrix of the first m singular vestors by Λ_m. Then a solution to the problem is determined by rescaling the normed singular vectors with formulas $Z_m = Z_m^* \sqrt{\Lambda_m}$ defining principal components, and $C_m = \sqrt{\Lambda_m} C_m^*$ defining their loadings.

Since singular vectors z corresponding to different singular values are mutually orthogonal, the factors can be found one by one as solutions to the one-factor model (5.2) above applied to the so-called residual data matrix: after a factor z and loadings c are found, X must be substituted by the matrix of residuals, $x_{il} \leftarrow x_{il} - c_l z_i$. The principal component of the residual matrix corresponds to the second largest singular value of the original matrix X. Repeating the process m times, one gets m first principal components and loading vectors.

It can be proven that, given m, the minimum value of $L = \sum_{i,l} e_{il}^2$ is equal to the sum of $r - m$ smallest singular values squared, $L(Z_m, C_m) = \sum_{k=m+1}^{r} \mu_k^2$, whereas m greatest singular values and corresponding singular vectors define the factor space solving the least-squares fitting problem for equation (5.3). Each $k - th$ component additively contributes μ_k^2 to the data scatter $T(X)$ $(k = 1, ..., m)$ so that equation (5.3), in the general case, becomes

$$T(X) = \mu_1^2 + ... + \mu_m^2 + L(Z_m, C_m).$$

Computation of singular vectors can be performed not necessarily with matrix X but with its derivative $M \times M$ matrix $X^T X$ or $N \times N$ matrix XX^T. The former can be derived by putting expression $z^* = Xc^*/\mu$ instead of z^* in the formula $X^T z^* = \mu c^*$ leading to $X^T X c^* = \mu^2 c^*$. This equation means that c^* is a latent vector of the square matrix $X^T X$ corresponding to its latent value μ^2. Thus, for $X^T X$, its latent value decomposition rather than SVD must be sought, because singular vectors of X are latent vectors of $X^T X$ corresponding to their latent values μ^2. Similarly, singular vectors z^* of X are latent vectors of XX^T.

In should be noted that in many texts the method of principal components is explained by using the square matrix $X^T X$ only, without any reference to the basic equations (5.2) or (5.4), see for instance [27, 61, 72]. The elements of matrix $X^T X$ are proportional to covariances or correlations between the variables; finding the maximum latent values and corresponding latent values of $X^T X$ can be interpreted as finding such a linear combination of the original variables that takes into account the maximum share of the data scatter. In this, the fact that the principal components are linear combinations of variables is an assumption of the method, not a corollary, which it is with models (5.2) and (5.4).

The singular value decomposition is frequently used as a data visualization tool on its own (see, for instance, [54]). Especially interesting results can be seen when entities are naturally mapped onto a visual image such as a geographic map; Cavalli-Sforza has interpreted several principal components in this way [13]. There are also popular data visualization techniques such as Correspondence analysis [6, 77], Latent semantic analysis [16], and Eigenfaces [133] that heavily rely on SVD. The former will be reviewed in the next section following the presentation in [93].

5.1.4 Correspondence factor analysis

Correspondence Analysis (CA) is a method for visually displaying both row and column categories of a contingency table $P = (p_{ij})$ $(i \in I, j \in J)$ in such a way that distances between the presenting points reflect the pattern of co-occurrences in P. There have been several equivalent approaches developed for introducing the method (see, for example, Benzécri [6]). Here we introduce CA in terms similar to those of PCA above.

To be specific, let us concentrate on the problem of finding just two "underlying" factors, $u_1 = \{(v_1(i)), (w_1(j))\}$ and $u_2 = \{(v_2(i)), (w_2(j))\}$, with $I \cup J$ as their domain, such that each row $i \in I$ is displayed as point $u(i) = (v_1(i), v_2(i))$ and each column $j \in J$ as point $u(j) = (w_1(j), w_2(j))$ on the plane as shown in Figure 5.1. The coordinate row-vectors, v_l, and column-vectors, w_l, constituting u_l $(l = 1, 2)$ are calculated to approximate the relative Quetelet coefficients $q_{ij} = p_{ij}/(p_{i+}p_{+j}) - 1$ according to equations:

$$q_{ij} = \mu_1 v_1(i) w_1(j) + \mu_2 v_2(i) w_2(j) + e_{ij} \tag{5.6}$$

where μ_1 and μ_2 are positive reals, by minimizing the weighted least-squares criterion

$$E^2 = \sum_{i\in I}\sum_{j\in J} p_{i+}p_{+j}e_{ij}^2 \tag{5.7}$$

with regard to μ_l, v_l, w_l, subject to conditions of weighted ortho-normality:

$$\sum_{i\in I} p_{i+}v_l(i)v_{l'}(i) = \sum_{j\in J} p_{+j}w_l(j)w_{l'}(j) = \begin{cases} 1, & l = l' \\ 0, & l \neq l' \end{cases} \tag{5.8}$$

where $l, l' = 1, 2$.

The weighted criterion E^2 is equivalent to the unweighted least-squares criterion L applied to the matrix with entries $a_{ij} = q_{ij}\sqrt{p_{i+}p_{+j}} = (p_{ij} - p_{i+}p_{+j})/\sqrt{p_{i+}p_{+j}}$. This implies that the factors are determined by the singular-value decomposition of matrix $A = (a_{ij})$. More explicitly, the optimal values μ_l and row-vectors $f_l = (v_l(i)\sqrt{p_{i+}})$ and column-vectors $g_l = (w_l(j)\sqrt{p_{+j}})$ are the maximal singular values and corresponding singular vectors of matrix A, defined by equations $Ag_l = \mu_l f_l$, $f_l A = \mu_l g_l$.

These equations, rewritten in terms of v_l and w_l, are considered to justify the joint display: the row-points appear to be averaged column-points and, vice versa, the column-points appear to be averaged versions of the row-points. The mutual location of the row-points is considered as justified by the fact that between-row-point squared Euclidean distances $d^2(u(i), u(i'))$ approximate chi-square distances between corresponding rows of the contingency table $\chi^2(i, i') = \sum_{j\in J} p_{+j}(q_{ij} - q_{i'j})^2$. Here $u(i) = (v_1(i), v_2(i))$ for v_1 and v_2 rescaled in such a way that their norms are equal to μ_1 and μ_2, respectively. To see it, one needs to derive first that the weighted averages $\sum_{i\in I} p_{i+}v_i$ and $\sum_{j\in J} p_{+j}w_j$ are equal to zero. Then, it will easily follow that the singularity equations for f and g are equivalent to equations $\sum_{j\in J} p(j/i)w_j = \mu v_i$ and $\sum_{i\in I} p(i/j)v_i = \mu w_j$ where $p(j/i) = p_{ij}/p_{i+}$ and $p(i/j) = p_{ij}/p_{+j}$ are conditional probabilities defined by the contingency table P. These latter equations define elements of v as weighted averages of elements of w, up to the factor μ, and vice versa.

The values μ_l^2 are latent values of matrix A^TA. As is known, the sum of all latent values of a matrix is equal to its trace, defined as the sum of diagonal entries, that is, $Tr(A^TA) = \sum_{t=1}^{r} \mu_t^2$ where r is the rank of A. On the other hand, direct calculation shows that the sum of diagonal entries of A^TA is

$$Tr(A^TA) = \sum_{i,j}(p_{ij} - p_{i+}p_{+j})^2/(p_{i+}p_{+j}) = X^2. \tag{5.9}$$

Thus,

$$X^2 = \mu_1^2 + \mu_2^2 + E^2 \tag{5.10}$$

Table 5.1: **Bribery:** Cross-classification of features Branch (X) and Type of Service (III) from Table 1.11.

Branch	Type of Service					Total
	ObstrJus	Favors	Extort	CategCh	Cover-up	
Government	0	8	7	0	3	18
LawEnforc	14	1	3	2	9	29
Other	1	1	0	5	1	8
Total	15	10	10	7	13	55

which can be seen as a decomposition of the contingency data scatter, measured by X^2, into contributions of the individual factors, μ_l^2, and unexplained residuals, E^2. (Here, $l = 1, 2$, but, actually, the number of factors sought can be raised up to the rank of matrix A.) In a common situation, the first two latent values account for a major part of X^2, thus justifying the use of the plane of the first two factors for visualization of the interrelations between I and J.

Thus, CA is analogous to PCA, but differing from PCA in the following aspects:

(i) CA applies to contingency data in such a way that relative Quetelet coefficients are modeled rather than original frequencies;

(ii) Rows and columns are assumed to have weights, the marginal frequencies, that are used in both the least-squares criterion and orthogonality equations;

(iii) Both rows and columns are visualized on the same display so that geometric distances between the representations reflect chi-square distances between row and column conditional frequency profiles;

(iv) The data scatter is measured by the Pearson chi-square association coefficient.

As shown in [6] (see also [77]), CA better reproduces the visual shapes of contingency data than the standard PCA.

Example 5.44. Contingency table for the synopsis of Bribery data and its visualization

Let us build on results of the generalization of the Bribery data set obtained by clustering in section 3.4.3: all features of the Bribery data in Table 1.12 are well represented by the interrelation between the two variables: the branch at which the corrupt service occurred and type of the service, features X and III in table 1.11, respectively. Let us take a look at the cross-classification of these features (Table 5.1) and visualize it with the method of Correspondence analysis.

On Figure 5.1, one can see which columns are attracted to which rows: Change of category to Other branch, Favors and Extortion to Government, and Cover-up and Obstruction of justice to Law Enforcement. This is compatible with the conclusions drawn in section 3.4.3. A unique feature of this display is that the branches constitute a triangle covering the services.

□

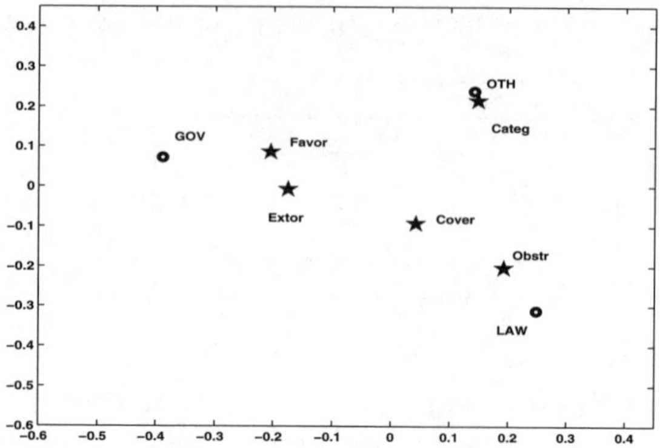

Figure 5.1: CA display for the rows and columns of Table 5.1 represented by circles and pentagrams, respectively.

5.2 Data recovery model for K-Means

5.2.1 Equation and data scatter decomposition

In K-Means, a clustering is represented by a partition $S = \{S_k\}$ of the entity set I consisting of K cluster lists S_k that do not overlap and cover all entities, that is, $S_k \cap S_l \neq \emptyset$ if $k \neq l$ and $\cup_{k=1}^{K} S_k = I$. The latter condition can be relaxed as described in sections 3.2.3 and 3.3. The lists are frequently referred to as cluster contents. The number of elements in S_k is frequently referred to as the cluster's size or cardinality and denoted by N_k. Centroids of clusters are vectors $c_k = (c_{kv})$ representing cluster "prototypes" or "standard" points.

Given a partition S and set of centroids $c = \{c_k\}$ resulting from K-Means, the original data can be recovered in such a way that any data entry y_{iv} (where $i \in I$ denotes an entity and $v \in V$ a category or quantitative feature) is represented by the corresponding centroid value c_{kv} such that $i \in S_k$, up to a residual, $e_{iv} = y_{iv} - c_{kv}$. In this way, clustering (S, c) leads to the data recovery model described by equations

$$y_{iv} = c_{kv} + e_{iv}, \; i \in S_k, \; k = 1, ..., K. \tag{5.11}$$

It is this model, in all its over-simplicity, that stands behind K-Means. Let us see how this may happen.

Multiplying equations (5.11) by themselves and summing up the results, it is not difficult to derive the following equation:

$$\sum_{i \in I} \sum_{v \in V} y_{iv}^2 = \sum_{v \in V} \sum_{k=1}^{K} N_k c_{kv}^2 + \sum_{i \in I} \sum_{v \in V} e_{iv}^2 \tag{5.12}$$

The derivation is based on the assumption that c_{kv} is the average of within cluster values y_{iv}, $i \in S_k$, so that $\sum_{i\in S_k} c_{kv}y_{iv} = N_k c_{kv}^2$. Noting that the right-hand term in (5.12), $\sum_{i\in I}\sum_{v\in V} e_{iv}^2 = \sum_{k=1}^{M}\sum_{i\in S_k}\sum_{v\in V}(y_{iv}-c_{kv})^2 = \sum_{k=1}^{M}\sum_{i\in S_k} d(y_i,c_k)$, is K-Means square error criterion $W(S,c)$ (3.2), equation (5.12) can be rewritten as

$$T(Y) = B(S,c) + W(S,c) \tag{5.13}$$

where $T(Y) = \sum_{i,v} y_{iv}^2$ is the data scatter, $W(S,c) = \sum_{k=1}^{M}\sum_{i\in S_k} d(y_i,c_k)$ square-error clustering criterion and $B(S,c)$ the middle term in decomposition (5.12):

$$B(S,c) = \sum_{v\in V}\sum_{k=1}^{K} c_{kv}^2 N_k \tag{5.14}$$

Equation (5.12), or its equivalent (5.13), is well-known in the analysis of variance; its parts $B(S,c)$ and $W(S,c)$ are conventionally referred to as between-group and within-group variance in statistics. In the context of model (5.11) these, however, denote the explained and unexplained parts of the data scatter, respectively. The square error criterion of K-Means, therefore, minimizes the unexplained part of the data scatter, or, equivalently, maximizes the explained part, $B(S,c)$ (5.14). In other words, this criterion expresses the idea of approximation of data Y by the K-Means partitioning as expressed in equation (5.11).

Equation (5.13) can be rewritten in terms of distances d since T(Y)=$\sum_{i=1}^{N} d(y_i,0)$ and $B(S,c) = \sum_{k=1}^{K} N_k d(c_k,0)$ according to the definition of the Euclidean distance squared:

$$\sum_{i=1}^{N} d(y_i,0) = \sum_{k=1}^{K} N_k d(c_k,0) + \sum_{k=1}^{K}\sum_{i=1}^{N} d(y_i,c_k) \tag{5.15}$$

5.2.2 Contributions of clusters, features, and individual entities

According to equations (5.13) and (5.14), each individual cluster $k = 1,...K$ additively contributes

$$B(S_k,c_k) = \sum_{v\in V} c_{kv}^2 N_k \tag{5.16}$$

to $B(S,c)$. In its turn, any individual cluster's contribution is the sum of cluster-specific feature contributions $B_{vk} = c_{kv}^2 N_k$. Following from the preliminary standardization of data by subtracting features' grand means, the contribution B_{vk} is proportional to the squared difference between variable v's grand mean c_v and its within cluster mean c_{kv}: the larger the difference the greater the contribution. This nicely fits into our intuition: the farther away is the cluster from the grand mean on a feature range, the more useful should be the feature in separating the cluster from the rest.

To evaluate contributions of individual entities to the explained part of the data scatter, one needs yet another reformulation of (5.16). Let us refer to the definition $c_{kv} = \sum_{i \in S_k} y_{iv}/N_k$ and put it into $c_{kv}^2 N_k$ "halfway." This then becomes $c_{kv}^2 N_k = \sum_{i \in S_k} y_{iv} c_{kv}$ leading to $B(S_k, c_k)$ reformulated as the summary inner product:

$$B(S_k, c_k) = \sum_{v \in V} \sum_{i \in S_k} y_{iv} c_{kv} = \sum_{i \in S_k} (y_i, c_k), \tag{5.17}$$

thus suggesting that the contribution of entity $i \in S_k$ to the explained part of the data scatter is (y_i, c_k), the inner product between the entity point and the cluster's centroid, as follows from (3.8). This may give further insights into the scatter decomposition to highlight contributions of individual entities (see an example in Table 3.26).

These and related interpretation aids are suggested for use in section 3.4.2 as a non-conventional but informative instrument.

5.2.3 Correlation ratio as contribution

To measure statistical association between a quantitative feature v and partition $S = \{S_1, ..., S_K\}$, the so-called correlation ratio η^2 has been defined in statistics:

$$\eta^2(S, v) = \frac{\sigma_v^2 - \sum_{k=1}^{K} p_k \sigma_{kv}^2}{\sigma_v^2} \tag{5.18}$$

where $\sigma_v^2 = \sum_{i \in I} (x_{iv} - c_v)^2/N$ and $\sigma_{kv}^2 = \sum_{i \in S_k} (x_{iv} - c_{kv})^2/N_k$ are the variance and within-cluster variance of variable v, respectively, before data pre-processing and $p_k = N_k/N$. Actually, the correlation ratio expresses the extent to which the within-cluster averages can be used as predicted values of v and, in this sense, is analogous to the determination coefficient in the model of linear regression. The correlation ratio ranges between 0 and 1, and it is equal to 1 only when all the within-class variances are zero. The greater the within-category variances, the smaller the correlation ratio.

Assuming that the raw data x_{iv} have been standardized to $y_{iv} = (x_{iv} - c_v)/b_v$ by shifting the origins to c_v and rescaling results by dividing them over b_v, it is not difficult to prove that the total contribution $B(S, v) = \sum_{k=1}^{K} B_{vk} = \sum_{k=1}^{K} c_{kv}^2 N_k$ of a quantitative feature v to the cluster-explained part of the data scatter in (5.14) equals to $N\eta^2(S, v)\sigma_v^2/b_v^2$ [94].

The cluster to variable contribution $N\eta^2(S,v)\sigma_v^2/b_v^2$ becomes plain $N\eta^2(S,v)$ when the variable has been normalized with b_v being its standard deviation, the option which hides the shape of the variable distribution. Otherwise, with b_v being the range r_v, the contribution should be considered a partition-to-feature association coefficient on its own:

$$\mu^2(S,v) = \eta^2(S,v)\sigma_v^2/r_v^2. \tag{5.19}$$

5.2.4 Partition contingency coefficients

Consider now the case of a nominal variable l presented by its set of categories V_l. The summary contribution $B(S,l) = \sum_{v\in V_l}\sum_k B_{vk}$ of nominal feature l to partition S, according to decomposition (5.12), appears to have something to do with association coefficients in contingency tables considered in section 2.2.3.

To analyze the case, let us initially derive frequency based reformulations of centroids for binary variables $v \in V_l$. Let us recall that a categorical variable l and cluster-based partition S, when cross classified, form contingency table p_{kv}, whose marginal frequencies are p_{k+} and p_{+v}, $k = 1, ..., K$, $v \in V_l$.

For any three-stage pre-processed column $v \in V_l$ and cluster S_k in the clustering (S,c), its within-cluster average is equal to:

$$c_{kv} = [\frac{p_{kv}}{p_{k+}} - c_v]/[b_v b_v^{'}] \tag{5.20}$$

where $b_v' = \sqrt{|V_l|}$. Indeed, within-cluster average in this case equals $c_{kv} = \frac{p_{kv}}{p_{k+}}$, the proportion of v in cluster S_k. The mean c_v of binary attribute $v \in V_l$ is the proportion of ones in it, that is, the frequency of the corresponding category, $c_v = p_{+v}$.

This implies that

$$B(S,l) = N\sum_{k=1}^{K} p_{k+} \sum_{v\in V_l} (p_{kv}/p_{k+} - p_{+v})^2/|V_l|b_v^2 \tag{5.21}$$

which can be transformed, with little arithmetic, into

$$B(S,l) = \frac{N}{|V_l|}\sum_{k=1}^{K}\sum_{v\in V_l}\frac{(p_{kv} - p_{k+}p_{+v})^2}{p_{k+}b_v^2} \tag{5.22}$$

where b_v and $|V_l|$ stand for the second stage, scaling, and the third stage, rescaling, during data pre-processing, respectively. The items summarized in (5.22) can be further specified depending on scaling coefficients b_v as

1. $\frac{(p_{kv}-p_k p_v)^2}{p_k}$ if $b_v = 1$, the range;

2. $\frac{(p_{kv}-p_k p_v)^2}{p_k p_v (1-p_v)}$ if $b_v = \sqrt{p_v(1-p_v)}$, Bernoulli standard deviation;

3. $\frac{(p_{kv}-p_k p_v)^2}{p_k p_v}$ if $b_u = \sqrt{p_u}$, Poisson standard deviation.

These lead to the following statement.

Statement 5.5. *The contribution $B(S,l)$ of a nominal variable l and partition S to the explained part of the data scatter, depending on the standardizing coefficients, is equal to contingency coefficient G^2 (2.11) or $Q^2 = X^2/N$ (2.12) if scaling coefficients b_v are taken to be $b_v = 1$ or $b_v = \sqrt{p_{+v}}$, respectively, and rescaling coefficients $b'_v = 1$. It is further divided by the number of categories $|V_l|$ if rescaling coefficients are $b'_v = \sqrt{|V_l|}$ for $v \in V_l$.*

Two well known normalizations of the Pearson chi-square contingency coefficient are due to Tchouprov, $T = X^2/\sqrt{(K-1)(|V_l|-1)}$, and Cramer, $C = X^2/\min(K-1, |V_l|-1)$, both symmetric over the numbers of categories and clusters. The statement 5.5. implies one more, asymmetric, normalization of X^2, $M = X^2/|V_l|$, as a meaningful part of the data scatter in the clustering problem.

When the chi-square contingency coefficient or related indexes are applied in the traditional statistics context, the presence of zeros in a contingency table becomes an issue because it contradicts the hypothesis of statistical independence. In the context of data recovery clustering, zeros are treated as any other numbers and create no problems at all because the coefficients are measures of contributions and bear no other statistical meaning in this context.

5.3 Data recovery models for Ward criterion

5.3.1 Data recovery models with cluster hierarchies

To formulate supporting models for agglomerative and divisive Ward clustering, one needs to explicitly define the concepts of cluster tree and hierarchy. A set $\mathbf{S}$ of subsets $S_w \subseteq I$ is called nested if for every two subsets S_w and $S_{w'}$ from $\mathbf{S}$ either one of them is part of the other or they do not overlap at all. Given a nested set $\mathbf{S}$, $S_w \in \mathbf{S}$ is referred to as a child of $S_{w'} \in \mathbf{S}$ if $S_w \subset S_{w'}$ and no other subset $S_{w''} \in \mathbf{S}$ exists such that $S_w \subset S_{w''} \subset S_{w'}$. A subset $S_w \in \mathbf{S}$ is referred to as a terminal node or leaf if S_w has no children in $\mathbf{S}$. A nested set $\mathbf{S}$ will be referred to as a cluster hierarchy over I if any non-terminal subset $S_w \in \mathbf{S}$ has two children $S_{w'}, S_{w''} \in \mathbf{S}$ covering it entirely so that $S_{w'} \cup S_{w''} = S_w$. The subsets $S_w \in \mathbf{S}$ will be referred to as clusters.

Two types of cluster hierarchy are of interest in modeling clustering algorithms: those $\mathbf{S}$ containing singleton clusters $\{i\}$ for all $i \in I$ and those containing set I itself as a cluster. The former will be referred to as the lower cluster hierarchy and the latter, the upper cluster hierarchy. A lower hierarchy can be thought of as resulting from an agglomerative clustering algorithm and an upper hierarchy from a divisive clustering algorithm. A complete result of a clustering algorithm of either type can be represented by a cluster tree, that is, a cluster hierarchy which is both lower and upper.

For an upper cluster hierarchy $\mathbf{S}$, let us denote the set of its leaves by $L(\mathbf{S})$; it is obviously a partition of I. Similarly, for a lower cluster hierarchy $\mathbf{S}$, let us denote its set of maximal clusters by $M(\mathbf{S})$; this is also a partition of I.

Given an upper or lower cluster hierarchy $\mathbf{S}$ over set I and a pre-processed data matrix $Y = (y_{iv})$, let us, for any feature v, denote the average value of y_{iv} within $S_w \in \mathbf{S}$ by c_{wv}.

Given an upper cluster hierarchy $\mathbf{S}$, let us consider its leaf partition $L(\mathbf{S})$. For any data entry y_{iv} and a leaf cluster $S_{w^*} \in L(\mathbf{S})$ containing it, the model underlying K-Means suggests that y_{iv} is equal to c_{w^*v} up to the residual $e_{iv} = y_{iv} - c_{w^*v}$. Obviously, $e_{iv} = 0$ if S_{w^*} is a singleton consisting of just one entity i. To extend this to the hierarchy $\mathbf{S}$, let us denote the set of all nonsingleton clusters containing i by $\mathbf{S}_i$ and add to and subtract from the equation the averages of feature v within each $S_w \in \mathbf{S}_i$. This leads us to the following equation:

$$y_{iv} = \sum_{S_w \in \mathbf{S}_i} (c_{w1,v} - c_{wv}) + e_{iv} \tag{5.23}$$

where S_{w1} is a child of S_w that runs through $\mathbf{S}_i$. Obviously, all $e_{iv} = 0$ in (5.23) if $\mathbf{S}$ is a cluster tree.

5.3.2 Covariances, variances and data scatter decomposed

Model in (5.23) is not just a trivial extension of the K-Means model to the case of upper cluster hierarchies, in spite of the fact that the only "real" item in the sum in (5.23) is c_{w^*v} where S_{w^*} is the leaf cluster containing i. Indeed, the equation implies the following decomposition.

Statement 5.6. *For every feature columns $v, u \in V$ in the pre-processed data matrix Y, their inner product can be decomposed over the cluster hierarchy $\mathbf{S}$ as follows:*

$$(y_v, y_u) = \sum_w \frac{N_{w1}N_{w2}}{N_w}(c_{w1,v} - c_{w2,v})(c_{w1,u} - c_{w2,u}) + (e_v, e_u) \tag{5.24}$$

where w runs over all nonterminal clusters $S_w \in \mathbf{S}$ with children S_{w1} and S_{w2}; N_w, N_{w1}, N_{w2} being their respective cardinalities.

Proof: The proof follows from equation $N_w c_w = N_{w1}c_{w1} + N_{w2}c_{w2}$ which relates the centroid of a non-terminal cluster S_w with those of its children. By putting this into (5.23), one can arrive at (5.24) by multiplying the decomposition for y_{iv} by that for y_{iu} and summing up results over all $i \in I$, q.e.d.

Another, more mathematically loaded analysis of model (5.23) can be based on the introduction of a $N \times m$ matrix Φ where N is the number of entities in I and m the number of nonterminal clusters in $\mathbf{S}$. The columns ϕ_w of Φ correspond to non-terminal clusters $S_w \in \mathbf{S}$ and are defined by equations: $\phi_{iw} = 0$ if $i \notin S_w$, $\phi_{iw} = a_w$ for $i \in S_{w1}$, and $\phi_{iw} = -b_w$ for $i \in S_{w2}$ where a_w and b_w are positive reals specified by the conditions that vector ϕ_w must be centered and normed. These conditions imply that $a_w = \sqrt{1/N_{w1} - 1/N_w} = \sqrt{N_{w2}/N_w N_{w1}}$ and $b_w = \sqrt{1/N_{w2} - 1/N_w} = \sqrt{N_{w1}/N_w N_{w2}}$ so that $a_w b_w = 1/N_w$. It is not difficult to prove that thus defined vectors ϕ_w are mutually orthogonal and, therefore, form an orthonormal base (see [90]). By using matrix Φ, equations (5.23) can be rewritten in matrix denotations as

$$Y = \Phi A + E \tag{5.25}$$

where A is a $m \times M$ matrix with entries $a_{wv} = \sqrt{N_{w1}N_{w2}/N_w}(c_{w1,v} - c_{w2,v})$. Multiplying (5.25) by Y^T on the left, one arrives at matrix equation $Y^T Y = C^T C + E^T E$, since $\Phi^T \Phi$ is the identity matrix and $\Phi^T E$ the zero matrix. This matrix equation is a matrix formulation for (5.24). This would give another proof of Statement 5.6.

Given a lower cluster hierarchy $\mathbf{S}$, model (5.23) remains valid, with e_{iv} redefined as $e_{iv} = c_{w\#i}$ where $S_{w\#} \in M(\mathbf{S})$ is the maximal cluster containing i. The summation in (5.23) still runs over all $S_w \in \mathbf{S}$ containing i, so that in the end the equation may be reduced to the definition $e_{hi} = c_{w\#i}$. Yet taken as they are, the equations lead to the same formula for decomposition of inner products between feature columns because of (5.25).

Statement 5.7. *Statement 5.6. is also true if* $\mathbf{S}$ *is a lower cluster hierarchy, with residuals redefined accordingly.*

These lead to a decomposition described in the following statement.

Statement 5.8. *Given a lower or upper cluster hierarchy* $\mathbf{S}$*, the data scatter can be decomposed as follows:*

$$\sum_{i \in I} \sum_{v \in V} y_{iv}^2 = \sum_{w} \frac{N_{w1}N_{w2}}{N_w} \sum_{v \in V} (c_{w1,v} - c_{w2,v})^2 + \sum_{i \in I} \sum_{v \in V} e_{iv}^2 \tag{5.26}$$

where w *runs through all nonterminal clusters* $S_w \in \mathbf{S}$.

Proof: To prove equation (5.26), set $u = v$ in equations (5.24) and sum them up over all feature columns $v \in V$. This also produces, on the left-hand side, the sum of all inner products (y_v, y_v), which is obviously the data scatter, and on the right side, exactly the right side of (5.26), q.e.d.

Note that the central sum in (5.26) is nothing but the squared Euclidean distance between centroids of clusters S_{w1} and S_{w2}, which leads to the following reformulation:

$$T(Y) = \sum_{w} \frac{N_{w1}N_{w2}}{N_w} d(c_{w1}, c_{w2}) + \sum_{i \in I} \sum_{v \in V} e_{iv}^2 \tag{5.27}$$

Further reformulations easily follow from the definitions of e_{iv} in upper or lower cluster hierarchies.

In particular, if **S** is a lower cluster hierarchy then the residual part $\sum_i \sum_v e_{iv}^2$ of the data scatter decomposition in (5.26) is equal to the complementary K-Means criterion $B(S, c)$ where $S = M(\mathbf{S})$ is the set of maximal clusters in **S** and c the set of their centroids. That means that for any lower cluster hierarchy **S** with $S = M(\mathbf{S})$:

$$T(Y) = \sum_{w} \frac{N_{w1}N_{w2}}{N_w} d(c_{w1}, c_{w2}) + B(S, c) \tag{5.28}$$

Similarly, if **S** is an upper cluster hierarchy, the residual part is equal to the original K-Means square error criterion $W(S, c)$ where $S = L(\mathbf{S})$ is the set of leaf clusters in **S** with c being their centroids. That means that for any upper cluster hierarchy **S** with $S = L(\mathbf{S})$:

$$T(Y) = \sum_{w} \frac{N_{w1}N_{w2}}{N_w} d(c_{w1}, c_{w2}) + W(S, c) \tag{5.29}$$

These decompositions explain what is going on in Ward clustering in terms of the underlying data recovery model. Every merging step in agglomerative clustering or every splitting step in divisive clustering adds the Ward distance

$$\lambda_w = dw(S_{w1}, S_{w2}) = \frac{N_{w1}N_{w2}}{N_w} d(c_{w1}, c_{w2}) \tag{5.30}$$

to the central sum in (5.27) by reducing the other part, $B(S, c)$ or $W(S, c)$, respectively. The central part appears to be that explained by the upper hierarchy in divisive clustering and that unexplained by the lower hierarchy in agglomerative clustering. Thus, the Ward distance (5.30) must be as large as possible in divisive clustering and as small as possible in agglomerative clustering.

The data recovery model for the K-Means partitioning leads to decomposition (5.13) of the data scatter and that for an upper cluster hierarchy (the divisive Ward-like clustering) to decomposition (5.29). By comparing these two equations, one can conclude that split contributions λ_w sum up to $B(S, c)$, that is, they give an alternative way of explaining clusters in S: by splits leading to S as the leaf set $L(\mathbf{S})$ of a cluster hierarchy rather than by clusters in S themselves. Obviously, decomposition (5.29) holds for any upper hierarchy leading to S, not necessarily that resulting from Ward-like clustering.

5.3.3 Direct proof of the equivalence between 2-Means and Ward criteria

On the first glance, the Ward criterion for dividing an entity set in two clusters has nothing to do with that of K-Means. Given $S_w \subseteq I$, the former is to maximize λ_w in (5.30) over all splits of S_w in two parts, while the K-Means criterion, with K=2, in the corresponding denotations is:

$$W(S_{w1}, S_{w2}, c_{w1}, c_{w2}) = \sum_{i \in S_{w1}} d(y_i, c_{w1}) + \sum_{i \in S_{w2}} d(y_i, c_{w2}) \tag{5.31}$$

Criterion (5.31) is supposed to be minimized over all possible partitions of S_w in two clusters, S_{w1} and S_{w2}. According to equation (5.15), this can be equivalently reformulated as the problem of maximization of:

$$B(S_{w1}, S_{w2}, c_{w1}, c_{w2}) = N_{w1} d(c_{w1}, 0) + N_{w2} d(c_{w2}, 0) \tag{5.32}$$

over all partitions S_{w1}, S_{w2} of S_w.

Now we are ready to prove that criteria (5.31) and (5.30) are equivalent.

Statement 5.9. *Maximizing Ward criterion (5.30) is equivalent to minimizing 2-Means criterion (5.31).*

Proof: To see if there is any relation between (5.30) and (5.32), let us consider an equation relating the two centroids with the total gravity center, c_w, in the entity set S_w under consideration:

$$N_{w1} c_{w1} + N_{w2} c_{w2} = (N_{w1} + N_{w2}) c_w \tag{5.33}$$

The equation holds because the same summary entity point stands on both sides of it.

Let us assume $c_w = 0$. This shouldn't cause any trouble because the split criterion (5.30) depends only on the difference between c_{w1} and c_{w2}, which does not depend on c_w. Indeed, if $c_w \neq 0$, then we can shift all entity points in S_w by subtracting c_w from each of them, thus defining $y'_{iv} = y_{iv} - c_{wv}$. With the shifted data, the averages are obviously, $c'_w = 0$, $c'_{w1} = c_{w1} - c_w$, and $c'_{w2} = c_{w2} - c_w$, which does not change the difference between centroids.

With $c_w = 0$, equation (5.33) implies $c_{w1} = (-N_{w2}/N_{w1})c_{w2}$ and $c_{w2} = (-N_{w1}/N_{w2})c_{w1}$. Based on these, the inner product (c_{w1}, c_{w2}) can be presented as either $(c_{w1}, c_{w2}) = (-N_{w2}/N_{w1})(c_{w1}, c_{w1})$ or $(c_{w1}, c_{w2}) = (-N_{w1}/N_{w2})(c_{w2}, c_{w2})$. By substituting these instead of (c_{w1}, c_{w2}) in decomposition $d(c_{w1}, c_{w2}) = [(c_{w1}, c_{w1}) - (c_{w1}, c_{w2})] + [(c_{w2}, c_{w2}) - (c_{w1}, c_{w2})]$ we can see that λ_w (5.30) becomes:

$$\lambda_w = \frac{N_{w1} N_{w2}}{N_w} [N_w (c_{w1}, c_{w1}) / N_{w2}] + [N_w (c_{w2}, c_{w2}) / N_{w1}].$$

Figure 5.2: Three clusters with respect to Ward clustering: which is the first to go, S_3 or S_1?

By removing redundant items, this leads to equation:

$$\lambda_w = B(S_{w1}, S_{w2}, c_{w1}, c_{w2}),$$

which completes the proof.

5.3.4 Gower's controversy

J. Gower [40] provided an example demonstrating a peculiarity of Ward distance as a clustering criterion to reflect the fact that factor $N_{w1}N_{w2}/(N_{w1} + N_{w2})$ in (4.1) or (5.30) favors the equal distribution of entities between split parts of a cluster and, thus, the criterion may fail to immediately separate singleton outliers.

To be more specific, let us refer to the following example (see Figure 5.2).

Let c_1, c_2, and c_3 be centroids of groups containing N_1, N_2, and N_3 entities respectively and all three located on a straight line so that c_2 lies between c_1 and c_3. The difference in distances would imply that, in a cluster hierarchy, c_1 and c_2 should be merged first with c_3 joining next, or, with splitting, c_3 would diverge first. However, this is not necessarily so with Ward distances.

Indeed, let the third group be quite small and consist, for instance, just of one entity. Let us further assume that the Euclidean squared distance d between c_2 and c_3 is 20 times as great as between c_2 and c_1. This means that $N_3 = 1$ and $d(c_2, c_3)/d(c_1, c_2) = 20$. Then the ratio of Ward distances $q = dw(c_2, c_3)/dw(c_1, c_2)$ will be equal to $q = 20(N_1 + N_2)/(N_1N_2 + 1)$. Depending on the value of q either of the mergers can be of favor according to the Ward criterion. If $q \geq 1$ then $dw(c_1, c_2) \leq d(c_2, c_3)$ and the intuitively obvious merger of S_1 and S_2 should go first. If, however, $q < 1$ then $dw(c_2, c_3)$ is smaller and S_2, S_3 are to be joined first. Obviously, q can be less than 1, for example, when $N_1 = N_2 = 50$, thus leading to first merging the more distant clusters with centroids c_2 and c_3.

Similarly, in Ward-like divisive clustering with this setting, the first split must be done to separate S_1, not the distant cluster S_3, from the rest.

This argumentation perhaps has contributed to the practice of using the unweighted distance between centroids $d(c_{w1}, c_{w2})$ as an ad hoc criterion in hierarchical clustering, the so-called UPGMA method [122, 28]. In our view, such a conclusion however would probably be an overstretching application of a mathematical property. In fact, with local search algorithms, which are the only ones currently available, the property may not work at all.

Let us consider, for instance, Ward-like divisive clustering with 2-Means splitting. It appears, the controversy does not show up in it. Indeed, one starts with the entities farthest from each other, c_1 and c_3, as initial seeds. Then, with the Minimum distance rule, all points in S_2 are put in the closest cluster S_1 and never removed, as the centroid of the merged cluster $S_1 \cup S_2$ is always closer to c_1 than c_3. This way, 2-Means will produce the intuitively correct separation of the farthest singleton from the rest.

Similarly, in Splitting by separating, c_3 is selected as the initial seed of the cluster to be separated. Then adding an element from the middle cluster will produce d/d' larger than the right-hand side expression in (4.7), thereby halting the splitting process. For instance, with $N_1 = N_2 = 50$ in the example above, $d/d' = 3.89$ while $(N_1N_2 + N_2)/(N_1N_2 - N_1) = 2.02$. Actually, the latter becomes larger than the former only after more than half of the entities from the middle cluster have been added to the distant singleton, which is impossible with local search heuristics.

This is an example of the situation in which the square error criterion would have led to a wrong partition if this was not prevented by constraints associated with the local search nature of 2-Means and Splitting by separating procedures.

5.4 Extensions to other data types

5.4.1 Similarity and attraction measures compatible with K-Means and Ward criteria

The number of different pair-wise similarity measures over multidimensional entities can be virtually infinite. The number of possible extensions of pair-wise similarity measures to cluster-by-cluster similarities can be quite large, too. The way of addressing these two issues is crucial for any clustering algorithm working with similarity data. We are going to use definitions that are, first, intuitively appealing, and, second, compatible with those implicitly involved in K-Means and Ward clustering.

The first issue to be addressed is of how to measure similarity a_{ij} between entities $i, j \in I$ provided that the raw data is in the format of an entity-to-feature table such as Table 1.10. When the data pre-processing is done as described in section 2.4, by shifting to grand means with follow-up scaling and rescaling, it is only natural to pick up the row-to-row inner product $a_{ij} = (y_i, y_j) = \sum_{v \in V} y_{iv}y_{jv}$ (note, the inner product, not the correlation coefficient) as a good similarity measure. Indeed, each feature v contributes product $y_{iv}y_{jv}$ to the total similarity, which points to the mutual locations of entities i and j on the axis v with respect to grand mean c_v because the data standardization translates c_v to the scale's origin, 0. The product is positive when both entities are either larger than $c_v = 0$ or smaller than $c_v = 0$. It is negative when i and j are on different sides from $c_v = 0$. These correspond to our intuition. Moreover, the closer y_{iv} and y_{jv} to 0, the smaller is the product; the further away is either of them from 0, the larger is the product. This property can be interpreted as supporting a major data mining idea that the less ordinary a phenomenon, the more interesting it is: "the ordinary" in this case is just the average.

The next issue to be addressed is measuring the quality of a cluster. Given a similarity matrix $A = (a_{ij})$, $i, j \in I$, let us introduce the following measure of the quality of any $S \subseteq I$:

$$A(S) = \sum_{i,j \in S} a_{ij}/N_S = N_S a(S) \tag{5.34}$$

where N_S is the number of entities in S and $a(S)$ its average internal similarity, $a(S) = \sum_{i,j \in S} a_{ij}/N_S^2$. The greater $A(S)$, the better is subset S as a cluster.

Why is this so? Conventionally, a cluster should be cohesive internally and separate externally. This is the leading thought in defining what is a good cluster in the literature. Criterion (5.34) does involve a measure of cohesion, the average similarity $a(S)$, but it seems to have nothing to do with measuring the separation of clusters from the rest. However, this is not exactly so. The right-hand expression in (5.34) shows that it is a compromise between two goals: maximizing the within cluster similarity $a(S)$ and maximizing the cluster's cardinality, N_S. The goals are at odds: the larger the cluster, the smaller its within cluster similarity. The factor N_S appears to be a proxy for the goal of making S separate from the rest, $I - S$. Let us formulate a property of the criterion supporting this claim.

For any entity $i \in I$, let us define its attraction to subset $S \subseteq S$ as the difference:

$$\beta(i, S) = a(i, S) - a(S)/2 \tag{5.35}$$

where: (a) $a(i, S)$ is the average similarity of i and S defined as $a(i, S) = \sum_{j \in S} a_{ij}/N_S$; (b) $a(S)$ is the average within S similarity. The fact, that $a(S)$ is equal to the average $a(i, S)$ over all $i \in S$, leads us to expect that normally $a(i, S) \geq a(S)/2$ for the majority of elements $i \in S$; that is, normally $\beta(i, S) \geq 0$ for $i \in S$ – entities should be positively attracted to their clusters. It appears, a cluster S maximizing the quality criterion $A(S)$ is much more than that: S is cohesive internally and separate externally because all its members are positively attracted to S whereas non-members are negatively attracted to S.

Statement 5.10. *If S maximizes $A(S)$ (5.34) and $N_S \geq 2$ then $\beta(i, S)$ is not negative for all $i \in S$ and not positive for all $i \notin S$.*

Proof: Indeed, if $i \in S$ and $\beta(i,S) < 0$, then $A(S-i) > A(S)$ which contradicts the assumption that S maximizes (5.34). To prove the inequality, let us consider $A(S)N_S = \sum_{i,j \in S} a_{ij}$, the sum of within cluster similarities. Obviously, $A(S)N_S = A(S-i)(N_S-1) + 2a(i,S)N_S - a_{ii}$. This leads to $[A(S-i) - A(S)](N_S - 1) = a_{ii} + A(S) - 2a(i,S)N_S = a_{ii} - 2\beta(i,S)N_S > 0$ since a_{ii} must be non-negative, which proves the inequality. The other part, that $i \notin S$ contradicts $\beta(i,S) > 0$, can be proven similarly, q.e.d.

Having thus discussed the intuition behind the similarity and cluster quality measures thus defined, it is nice to see that they have something to do with K-Means and Ward criteria.

Statement 5.11. *If the similarity measure a_{ij} is defined as the inner product (y_i, y_j), then, for any partition $S = \{S_1, ..., S_K\}$ with the set of centroids $c = \{c_1, ..., c_K\}$, the relative within-cluster similarity measure $A(S_k)$ is equal to the contribution $N_k \sum_{v \in V} c_{kv}^2$ of S_k to the complementary cluster criterion $B(S,c)$ in decomposition (5.13):*

$$A(S_k) = \sum_{i,j \in S_k} a_{ij}/N_k = N_k \sum_{v \in V} c_{kv}^2 \tag{5.36}$$

Proof: Indeed, according to definition, $c_{kv} = \sum_{i \in S_k} y_{iv}/N_k$, which implies that $c_{kv}^2 = \sum_{i,j \in S_k} y_{iv}y_{jv}/N_k^2$. Summing these up by $v \in V$, we get (5.36), q.e.d.

This statement leads to the following.

Statement 5.12. *Criteria maximized by K-Means partitioning and the divisive Ward-like method can be expressed in terms of between entity inner products $a_{ij} = (y_i, y_j)$ as*

$$B(S,c) = \sum_{k=1}^{K} A(S_k) = \sum_{k=1}^{K} \sum_{i,j \in S_k} a_{ij}/N_k \tag{5.37}$$

and

$$\lambda_w = A(S_{w1}) + A(S_{w2}) - A(S_w) \tag{5.38}$$

where items on the right are defined in (5.34).

Expression (5.38) actually follows from the property of Ward distance (4.2) and decomposition (5.13). Since the right-hand part in it does not depend on the split to be made in S_w and thus can be omitted, the remaining part is nothing but partition criterion (5.37) adjusted to the case of bi-class partitioning.

Obviously, criteria (5.37) and (5.38) can be applied to any similarity measure a_{ij}, not necessarily the inner product.

A method for building clusters individually, one at a time, according to criterion (5.37), will be described in section 5.5.5. In this section we present only a method for divisive clustering using similarity data with criterion (5.38).

The sum $\sum_{i,j \in S_w} a_{ij}$ in the parent class S_w is equal to the sum of within children summary similarities plus the between children similarity doubled. Thus, by subtracting the former from their counterparts in criterion (5.38), the criterion can be transformed into:

$$\lambda_w = \frac{N_{w1}N_{w2}}{N_w}(A_{11} + A_{22} - 2A_{12}) \qquad (5.39)$$

where

$$A_{ij} = \frac{\sum_{i \in S_{wi}} \sum_{j \in S_{wj}} a_{ij}}{N_{wi}N_{wj}},$$

the average similarity between S_{wi} and S_{wj} or within S_{wi} if $i = j$ $(i, j = 1, 2)$.

Note that criterion (5.39) is but the Ward distance translated in terms of entity-to-entity inner products treated as similarities. Moving an entity i from S_{w2} to S_{w1} leads to the change in λ_w equal to

$$\delta(i) = A_{22}\frac{N_{w2}}{N_{w2} - 1} - A_{11}\frac{N_{w1}}{N_{w1} + 1} + 2(A1(i, w1) - A2(i, w2)) + \alpha A_{ii} \qquad (5.40)$$

where $A1(i, w1) = \sum_{j \in S_{w1}} a_{ij}/(N_{w1} + 1)$, $A2(i, w2) = \sum_{j \in S_{w2}} a_{ij}/(N_{w2} - 1)$ and $\alpha = N/((N_{w1} + 1)(N_{w2} - 1))$. This can be derived from expression (5.38). A similar value for the change in λ_w when $i \in S_{w1}$ is moved into S_{w2} can be obtained from this by interchanging symbols 1 and 2 in the indices.

Let us describe a local search algorithm for maximizing criterion λ_w in (5.39) by splitting S_w in two parts. At first, a tentative split of S_w is carried out according to the dissimilarities $\lambda(i, j) = (a_{ii} + a_{jj} - 2a_{ij})/2$ that are defined according to the criterion (5.39) applied to individual entities. Then the split parts are updated by exchanging entities between them until the increment δ in (5.40) becomes negative.

Splitting by similarity

1. *Initial seeds.* Find a pair of entities i^*, j^* maximizing $\lambda(i, j) = (a_{ii} + a_{jj} - 2a_{ij})/2$ over $i, j \in S_w$.
2. *Initial clusters.* Create initial S_{w1} and S_{w2} by putting i^* into the former and j^* into the latter and by distributing each $i \in S_w$ either to S_{w1} or S_{w2}, according to the sign of an analogue to (5.40), $\delta(i) = 3(a_{ii^*} - a_{ij^*}) - (a_{i^*i^*} - a_{j^*j^*})$.
3. *Candidacy marking.* For any $i \in S_w$ calculate $\delta(i)$ and take i^* maximizing $\delta(i)$.
4. *Exchange.* If $\delta(i^*) > 0$, move i^* to the other split part and go to step 3. Otherwise, halt the process and output current clusters S_{w1} and S_{w2} along with corresponding λ_w.

Table 5.2: Attractions of Masterpieces to hierarchical clusters.

	Masterpieces							
Cluster	1	2	3	4	5	6	7	8
1,2,3,4,5,6	0.12	0.04	0.05	0.02	-0.05	0.07	-0.24	-0.36
7,8	-0.86	-0.63	-0.67	-0.58	-0.37	-0.71	**0.20**	**0.56**
1,2,3	**0.29**	**0.28**	**0.15**	-0.39	-0.52	-0.57	-0.61	-0.54
4,5,6	-0.39	-0.54	-0.38	**0.10**	**0.09**	**0.37**	-0.20	-0.51

In this method, step 3 is the most challenging computationally: it requires the finding of a maximum $\delta(i)$ over $i \in I$. To scale the method to large data sizes, one can use not the optimal i but rather any i at which $\delta(i) > 0$: this would be analogous to shifting from the method of steepest descent to the method of possible directions in the optimization of a function.

In spite of its simplicity, the Splitting by similarity produces rather tight clusters, which can be expressed in terms of the measure of *attraction* $\beta(i, S) = a(i, S) - a(S)/2$ (5.35).

If an entity $i \in S_w$ is moved from split part S_{w2} to S_{w1}, the change in the numbers of elements of individual parts can be characterized by the number $n_{12} = \frac{Nw1(N_{w2}-1)}{N_{w2}(N_{w1}+1)}$ or by the similar number n_{21} if an entity is moved in the opposite direction. Obviously, for large N_{w1} and N_{w2}, n_{12} tends to unity. It appears that elements of a split part are more attracted to it than to the other part, up to this quantity.

Statement 5.13. *For any* $i \in S_{w2}$,

$$n_{12}\beta(i, S_{w1}) < \beta(i, S_{w2}) \tag{5.41}$$

and a symmetric inequality holds for any $i \in S_{w1}$.

Proof: Indeed, $\delta(i) < 0$ for all $i \in S_{w2}$ in (5.40) after the Splitting by similarity algorithm has been applied to S_w. This implies that $\frac{N_{w1}}{N_{w1}+1}\beta(i, S_{w1}) + \alpha < \frac{N_{w2}}{N_{w2}-1}\beta(i, S_{w2})$. The inequality remains true even if α is removed from it because $\alpha > 0$. Dividing the result by $\frac{N_{w2}}{N_{w2}-1}$ leads to (5.41), q.e.d.

A divisive clustering algorithm can be formulated exactly as in section 4.2 except that this time Splitting by similarity is to be used for splitting clusters.

Example 5.45. Attractions of entities to Masterpieces clusters

Table 5.2 displays the attractions of entities in the Masterpieces data to clusters occurring at the splitting process. In the end, not only Statement 5.13. holds but all entities become positively attracted to their clusters and negatively to the rest.

□

5.4.2 Application to binary data

The case at which all variables are binary is of considerable interest because it emerges in many application areas such as document clustering in which features characterize presence/absence of keywords. When keywords are created automatically from collections of documents, the size of the feature space may become much larger than the size of the entity set so that it can be of advantage to shift from the entity-to-feature data format to a less demanding search space such as similarities between entities or just word counts. The K-Means and Ward criteria can be adjusted for use in these situations.

Inner product expressed through frequencies

As stated in the previous section, the inner product of rows in the pre-standardized data matrix is a similarity measure which is compatible with K-Means and Ward criteria. With binary features, specifying $c_v = p_v$, the proportion of entities at which feature v is present, and $b_v = 1$, the range, the inner product can be expressed as:

$$a_{ij} = \sum_{v \in V} y_{iv} y_{jv} = |V_i \cap V_j| - |V_i| - |V_j| + t(i) + t(j) + \gamma \tag{5.42}$$

where $t(i)$ (or $t(j)$) is the total frequency weight of features that are present at i (or, at j). The frequency weight of a feature v is defined here as $1 - p_v$; the more frequent is the feature the less its frequency weight. The value of $\gamma = \sum_v p_v^2$ here is simply an averaging constant related to the Gini coefficient. The right-hand expression for a_{ij} follows from the fact that each of pre-processed y_{iv} and y_{jv} can only be either $1 - p_v$ or $-p_v$.

Similarity index (5.42) is further discussed in section 7.3.1. Here we note that it involves evaluations of the information contents of all binary features that are present at entities i or j.

Binary case criteria

With binary features, K-Means and Ward clustering criteria can be reformulated in terms of feature counts that can lead to scalable heuristics for their optimization.

Indeed, given a partition $S = \{S_k\}$ along with cluster centroids $c_k = (c_{kv})$, and standardizing coefficients $a_v = p_v$ and b_v, the within cluster average values of binary features can be expressed as $c_{kv} = [(1-p_v)p_{kv}/p_k - p_v(1-p_{kv}/p_k)]/b_v$ where p_{kv} is the proportion of entities simultaneously falling in feature v and cluster S_k, and p_k is the proportion of S_k in I. This leads to

$$c_{kv} = (p_{kv}/p_k - p_v)/b_v. \tag{5.43}$$

Putting equation (5.43) into the formulas for K-Means criterion and Ward distance proves the following statement.

Statement 5.14. *In the situation of binary data pre-processed with $a_v = p_v$, the complementary criterion $B(S,c)$ of K-Means is equal to*

$$B(S,c) = N\sum_{k=1}^{K}\sum_{v\in V}\frac{(p_{kv} - p_k p_v)^2}{p_k b_v^2} \tag{5.44}$$

and Ward distance to

$$\lambda_w = \frac{N_{w1}N_{w2}}{N_w}\sum_{v\in V}\left(\frac{p_{w1,v}}{p_{w1}} - \frac{p_{w2,v}}{p_{w2}}\right)^2/b_v^2 \tag{5.45}$$

It should be noted that, when $b_v = 1$, measure λ_w (5.45) closely resembles the so-called "twoing rule," a heuristic criterion used in CART techniques for scoring splits in decision trees [11] and $B(S,c)$ (5.44) is the sum of NG^2 over all nominal features involved, where G^2 is Quetelet coefficient (2.11) (and (2.16)) introduced in section 2.2.3.

5.4.3 Agglomeration and aggregation of contingency data

Specifics of contingency and flow data have been pointed out in sections 2.5.2, 4.4.2 and 2.2.3. Basically, they amount to the highest uniformity of all data entries so that they can be meaningfully summed up and naturally pre-processed into Quetelet association indexes. These specifics are dwelt on in this section.

Agglomeration

Let $P(T,U) = (p_{tu})$ be a contingency matrix with rows and columns representing categories $t \in T$ and $u \in U$, respectively, and $F = \{F_1, ..., F_m\}$ a partition of row set T in non-overlapping nonempty classes F_1, ...,F_m. Since entries p_{tu} can be meaningfully summed up all over the table, a partition F can be used not only for clustering but also for aggregating matrix $P(T,U)$ into $P(F,U) = (p_{fu})$ $(f = 1, ..., m)$ where $p_{fu} = \sum_{t\in F_f} p_{tu}$ is the proportion of entities co-occurring in class F_f and category u for any $f = 1, ..., m$ and $u \in U$.

The row-to-column associations in $P(T,U)$ and $P(F,U)$ can be represented with relative Quetelet coefficients $q_{tu} = p_{tu}/(p_{t+}p_{+u}) - 1$ and $q_{fu} = p_{fu}/(p_{f+}p_{+u}) - 1$ defined in section 2.2.3. The match between the aggregate indexes q_{fu} and the individual entry indexes q_{tu}, $t \in F_f$, can be scored with an aggregate weighted square error index, $L(F,U) = \sum_{t\in T}\sum_{u\in U} p_{t+}p_{+u}e_{tu}^2$: the smaller L, the better $P(F,U)$ aggregates $P(T,U)$.

To support the choice of $L(F,U)$ as an aggregation criterion, the following equation can be derived with elementary algebraic manipulations: $L(F,U) =$

$X^2(T,U) - X^2(F,U)$ where X^2 is the Pearson contingency coefficient in the format of equation (2.13) applied to $P(T,U)$ and $P(F,U)$, respectively. This equation can be put as:

$$X^2(T,U) = X^2(F,U) + L(F,U), \tag{5.46}$$

and interpreted as a Pythagorean decomposition of the data scatter: $X^2(T,U)$ is the original data scatter, $X^2(F,U)$ is its part taken into account by partition F, and $L(F,U)$ is the unexplained part. Obviously, minimizing $L(F,U)$ over partition F is equivalent to maximizing $X^2(F,U)$.

Let $F(k,l)$ be the partition obtained from F by merging its classes F_k and F_l into united class $F_{k\cup l} = F_k \cup F_l$. To score the quality of the merging, we need to analyze the difference $D = X^2(F,U) - X^2(F(k,l),U)$, according to (5.46). Obviously, D depends only on items related to k and l:

$$D = \sum_{u\in U} [p_{ku}q_{ku} + p_{lu}q_{lu} - (p_{ku} + p_{lu})q_{k\cup l,u}] =$$

$$\sum_{u\in U} (1/p_{+u})[p_{ku}^2/p_{k+} + p_{lu}^2/p_{l+} - (p_{ku} + p_{lu})^2/(p_{k+} + p_{l+})].$$

By using equation $(x-y)^2 = x^2 + y^2 - 2xy$, one can transform the expression on the right to obtain:

$$D = \sum_{u\in U} (1/p_{+u}) \frac{p_{ku}^2 p_{l+}^2 + p_{lu}^2 p_{k+}^2 - 2p_{ku}p_{lu}p_{k+}p_{l+}}{p_{k+}p_{l+}(p_{k+} + p_{l+})},$$

which leads to:

$$D = \frac{p_{k+}p_{l+}}{p_{k+} + p_{l+}} \sum_{u\in U} (1/p_{+u})(p_{ku}/p_{k+} - p_{lu}/p_{l+})^2.$$

This is exactly the chi-square distance $\chi(F_k, F_l)$ considered in sections 2.5.2 and 4.4.2.

Thus, it is proven that the chi-square distance represents the increase in criterion $L(F,U)$ when two clusters of F are merged. In this aspect, it parallels the conventional Ward distance manifested in the criterion λ_w (5.30).

In fact, all the clustering theory in sections 5.2 and 5.3 can be extended to the contingency data case via equation (5.46). Aggregation of columns can be included into the analysis as well, as described in [90], pp. 323-327.

Similar considerations can be applied to the case of an interaction table, that is, a flow data table in which the row and column sets coincide, $T = U$ as, for instance, in the Digit confusion data. The agglomeration procedure remains the same as in the case when only rows are aggregated. However, since aggregation of rows here will also involve columns, the change of the Pearson chi-square coefficient cannot be expressed with chi-square distances between rows or columns, which makes computations much more complicated.

5.4.4 Extension to multiple data

The power of the data recovery approach can be further demonstrated in problems involving data on the same entities coming from different sources in different formats. For instance, two data tables on numeral digits, Digits Table 1.8 of their drawing features and Confusion Table 1.9, are given in section 1.1. We have analyzed these data by finding patterns in one of them and using the other one for interpretation of the patterns. However, in some applications one may need to look for patterns that are present in both data sets. Obviously, to do this, data sets must have a common entity set, which is so in the example of Digits. Problems of this type are of interest in such applications as data fusion.

Although generally many different data sets involving the same entities can be available, we will concentrate here on the data formats that are given for Digits. Thus, the case is analyzed in which an entity set I is provided with two data sets, a pre-processed standardized entity-to-feature data matrix Y and a flow table $P(I,I) = P = (p_{ij})$, $i,j \in I$. These two data sets can be of different importance in description of the phenomenon under consideration. To take this into account, a relative weight $\alpha > 0$ can be assigned to Y with $P(I,I)$ assumed to have weight 1.

Let us consider a partition $S = \{S_1, ..., S_K\}$ of I with classes S_k corresponding to the data patterns that are being searched for. Let us denote the binary membership vector for S_k by $z_k = (z_{ik})$ so that $z_{ik} = 1$ if $i \in S_k$ and $z_{ik} = 0$ if $i \notin S_k$. According to the data recovery approach, this partition can be used to recover the feature data with an analogue to the K-Means clustering model (5.11) which indeed can be presented as an extension of the Principal component analysis model (5.4) (this is discussed in a greater detail later in section 5.5.3, see (5.52)):

$$y_{iv} = c_{1v}z_{i1} + c_{2v}z_{i2} + ... + c_{Kv}z_{iK} + e_{iv}, \tag{5.47}$$

for all $i \in I$ and $v \in V$, and the flow data with a similar equation, though taking into account that the row and column entities are the same:

$$q_{ij} = \sum_{k=1}^{K}\sum_{l=1}^{K} \lambda_{kl} z_{ik} z_{jl} + \epsilon_{ij}, \tag{5.48}$$

for all $i,j \in I$, where c_{kv} and λ_{kl} are unknown reals.

The partition S and reals c_{kv} and λ_{kl} are sought to minimize the least squares criterion:

$$L = \alpha \sum_{i\in I}\sum_{v\in V} e_{iv}^2 + \sum_{i,j\in I} p_{i+}p_{+j}\epsilon_{ij}^2 \tag{5.49}$$

in which coefficient α reflects the prior importance of Y relative to that of P.

To develop a K-Means-like method of alternating minimization of L, let us consider S fixed. Then the optimal c_{kv} is obviously the average of y_{iv} over $i \in S_k$, and the optimal λ_{kl} is the Quetelet coefficient q_{kl} calculated over the aggregate interaction matrix $P(S,S)$, entries of which are defined as $p_{kl} = \sum_{i\in S_k}\sum_{j\in S_l} p_{ij}$, $k,l = 1,...,K$.

Then, given cluster centroids $c_k = (c_{kv})$ and $\lambda_{kl} = q_{kl}$, criterion L (5.49) can be rewritten as

$$L = \sum_{k=1}^{K}\sum_{i\in S_k}[\alpha d(y_i, c_k) + p_{i+}\chi(q_i, q(S))] \tag{5.50}$$

where d and χ are the squared Euclidean and chi-square distances, respectively. For $i \in S_k$, $\chi(q_i, q(S)) = \sum_{l=1}^{K}\sum_{j\in S_l} p_{+j}(q_{ij} - q_{kl})^2$, so that the expression in square brackets,

$$D(i,k) = \alpha d(y_i, c_k) + p_{i+}\chi(q_i, q(S)), \tag{5.51}$$

expresses the contribution of $i \in S_k$ to L. Obviously, to minimize (5.50) with regard to all possible assignments of entities to classes S_k, an extension of the Minimum distance rule should be utilized at which $D(i,k)$ plays the role of distance $d(y_i, c_k)$ in K-Means: each $i \in I$ should be assigned to such S_k which minimizes $D(i,k)$ over $k = 1,...,K$. The role of the weight α is clearly seen in (5.51): the greater the α the greater the contribution of distance $d(y_i, c_k)$ relative to the other item coming in $D(i,k)$ from P.

Now one can formulate a method replicating K-Means for the situation of two data sets.

Straight Fusion K-Means

1. *Initial setting.* Choose a candidate partition S of I.
2. *Parameter update.* Given S, calculate centroids c_k of S_k and aggregate confusion table $P(S,S)$ with aggregate Quetelet coefficients q_{kl}.
3. *Partition update.* Given centroids and aggregate Quetelet coefficients, calculate $D(i,k)$ (5.51) for all $i \in I$ and $k = 1,...,K$, and assign each i to that S_k at which $D(i,k)$ is minimum.
4. *Stop condition.* Check whether the new partition coincides with the previous one. If yes, halt; otherwise, go to Step 2.

The process eventually stabilizes because the number of partitions is finite and criterion L (5.50) decreases at each stage.

To formalize the issue of choosing the prior importance coefficient α, one needs to define a more general clustering criterion, an example of which can be found in [100].

5.5 One-by-one clustering

5.5.1 PCA and data recovery clustering

As described in section 5.1.3, the Principal component analysis (PCA) is a method for extracting orthogonal factors from data one by one. This strategy is well supported by the existence of the singular-value decomposition SVD of matrices.

Recently, some efforts have been made to apply PCA to both partitioning [49] and divisive clustering [10]. The efforts are based on consecutively executing a two-step procedure at which a principal component is found at the first step, and it is used for clustering at the second step. Data recovery models described above lead to different extensions of PCA to clustering. We do not apply PCA to clustering but rather modify PCA by using the same data recovery model adjusted to the case of clustering by specifying certain constraints: the factor score vectors are required to be binary in partitioning, or ternary in hierarchical clustering. This strategy can be followed because of striking similarities between the data recovery models for PCA and clustering.

As explained in section 5.1.3, the PCA model assumes a small number m of hidden factors, the principal components, z_1, z_2, ..., z_m, to underlie the observed matrix Y so that $Y = ZA + E$ where Z is an $N \times m$ matrix whose columns are the principal components, A is a $m \times M$ matrix of so-called factor loadings and E is a matrix of residuals that are to be least-square minimized with respect to the sought Z and A. Each optimal component z_t additively contributes μ_t^2 to the explained part of the data scatter $T(Y)$ $(t = 1, ..., m)$. Here μ_t is a singular value of Y or, equivalently, μ_t^2 is a latent value of both feature-to-feature covariance matrix Y^TY and entity-to-entity similarity matrix YY^T. The principal components can be found one by one in such a way that the PCA model with $m = 1$ is iteratively applied to the residual data matrix Y from which principal components found at previous steps have been subtracted. The first iteration yields the component corresponding to the maximum singular value; the second yields the next largest singular value, etc.

The data recovery models in (5.11), for partitioning, and (5.23), for hierarchical clustering, also have the format of equation $Y = ZA + E$ with residuals E to be least-square minimized over unknown Z and A. In the hierarchical clustering problem Z is a matrix Φ of ternary vectors ϕ_w corresponding to non-terminal clusters S_w of a lower or upper cluster hierarchy to be built, as explained on page 154, see equation (5.25). In the partitioning problem, Z is a matrix of the cluster membership vectors z_k defined by clusters S_k so that $z_{ik} = 1$ for $i \in S_k$ and $z_{ik} = 0$ for $i \notin S_k$. Thus defined, vectors z_k are mutually orthogonal, because clusters S_k do not overlap.

The model for partition clustering with criterion (3.2) is equivalent to the PCA model with $m = K$ and the constraint that Z must be a binary matrix of the cluster membership vectors. Similarly, the model for hierarchical clustering is equivalent to the PCA model with the restriction that Z must be a matrix of m ternary split membership vectors. The PCA strategy of extracting columns of Z one by one, explained in section 5.1.3, thus can be applied in each of these situations. We begin by describing an adaptation of this strategy to the case of hierarchical clustering and then to the case of partitioning. Extensions to the similarity and contingency data are briefly described next (for more detail, see [90]).

5.5.2 Divisive Ward-like clustering

Given an upper cluster hierarchy $\mathbf{S} = \{S_w\}$, each item $c_{w1} - c_w$ in equation (5.23) with c_{w1} being the centroid of $S_{w1} \in \mathbf{S}$ and c_w the centroid of the parent S_w, contributes $\lambda_w = wd(S_{w1}, S_{w2}) = \frac{N_{w1}N_{w2}}{N_w} d(c_{w1}, c_{w2})$ (5.30) to data scatter $T(Y)$ according to equation (5.29). This implies that, to build $\mathbf{S} = \{S_w\}$ from scratch, that is, by starting from the universal cluster $S_w = I$, one can employ the PCA one-by-one strategy by adding to $\mathbf{S}$ one split at a time, each time maximizing contribution λ_w of the split to the explained part of the data scatter so that λ_w plays the role of the squared singular value μ_t^2 of Y. The analogy is not superficial. Both μ_t and $\sqrt{\lambda_w}$ express the scaling coefficient when projecting the data onto a corresponding axis which follows the principal component z_t, for the former, and the ternary split vector ϕ_w, for the latter.

In the sequel to this section, we are going to demonstrate that the divisive Ward-like algorithm described in section 4.2 is an implementation of the PCA one-by-one extracting strategy. The strategy involves two types of iterations. One, major iteration, builds a series of splits of clusters S_w into split parts S_{w1} and S_{w2} to greedily maximize the summary contribution $\sum_{S_w \in \mathbf{S}} \lambda_w$ to the data scatter $T(Y)$. The other, minor iteration, is utilized when maximizing individual λ_w by splitting S_w using a specific splitting method. Two such methods described in section 4.2 fit into the strategy because they do maximize λ_w, though locally. They are: a straight method, with 2-Means, and an incremental method, Splitting by separation, based on a different expression, (4.6), for λ_w.

The use of 2-Means with Euclidean squared distance d, not Ward distance dw, for splitting is justified with the following statement.

Statement 5.15. *The 2-Means algorithm for splitting S_w maximizes (locally) the Ward distance between split parts.*

Proof: According to equation (5.29), the criterion minimized in splitting of partition $L(\mathbf{S})$ for an upper cluster hierarchy $\mathbf{S}$ by maximizing the Ward distance (5.30) is $W(L(S), c(L(S)))$, the K-Means criterion, q.e.d.

Another proof of this statement is given in section 5.3.3.

Each splitting step is equivalent to finding a ternary splitting vector maximizing Ward distance between its parts or, equivalently, minimizing the summary squared residuals in (5.27) with regard to all possible tertiary column vectors $\phi_w = (\phi_{iw})$ and any row-vectors $a_w = (a_{wv})$ in model (5.25). According to the first-order optimality conditions, the optimal vector a_w in (5.25) relates to the ϕ_w as the weighted vector of differences, $a_{wv} = \sqrt{N_{w1}N_{w2}/N_w}(c_{w1,v} - c_{w2,v})$, which concurs with the data recovery model (5.23) for hierarchical clustering.

5.5.3 Iterated Anomalous pattern

In this section we are going to demonstrate that the PCA one-by-one strategy applied when the membership vectors are constrained to be binary, is but the iterated Anomalous pattern analysis.

Consider a set of subsets $S_k \subseteq I$, each specified by a binary N-dimensional vector $z_k = (z_{ik})$ where $z_{ik} = 1$ if $i \in S_k$ and $z_{ik} = 0$ if $i \notin S_k$. (Vector z_k is interpreted as the membership vector for subset $S_k \subseteq I$.) The data recovery model (5.11) for K-Means can be rewritten with z_k in the following way:

$$y_{iv} = c_{1v}z_{i1} + c_{2v}z_{i2} + ... + c_{Kv}z_{iK} + e_{iv}, \tag{5.52}$$

which follows the PCA model (5.4) with $m = K$ factors and the additional requirement that scoring vectors z_k must be binary.

The one-by-one extracting strategy here would require the building of clusters S_k one by one, each time minimizing criterion:

$$l = \sum_{i\in I}\sum_{v\in V}(y_{iv} - c_v z_i)^2 \tag{5.53}$$

over unknown c_v and binary z_i, index k being omitted. The membership vector z is characterized by subset $S = \{i : z_i = 1\}$. Criterion (5.53) can be rewritten in terms of S, as follows:

$$W(S, c) = \sum_{i\in S} d(y_i, c) + \sum_{i\notin S} d(y_i, 0) \tag{5.54}$$

This is the square error criterion (3.2) for a partition consisting of two clusters, S and its complement $\bar{S} = I - S$, with regard to their respective centroids, c and 0. However, in contrast to the K-Means method, the centroid 0 here is not the center of gravity of the complementary set $I - S$, but is kept constant and does not change when S and $I - S$ change.

Given S, the optimal c in (5.54) is obviously the center of gravity of S because the first sum is constant there. Given c, a subset S to minimize (5.54) must include every $i \in I$ such that $d(y_i, c) < d(y_i, 0)$. Obviously, these are exactly the rules in Steps 3 and 4 of the Anomalous pattern clustering algorithm in section 3.2.3.

Thus, the following is proven.

Statement 5.16. *The Anomalous pattern algorithm iteratively minimizes (5.54) according to the alternating optimization rule: given c, find optimal S, and given S, find optimal c. The initial choice of c in the Anomalous cluster minimizes (5.54) over all singleton clusters.*

Thus, the AP method is a set of minor iterations of alternatingly optimizing criterion (5.53) for selecting the best cluster.

After one cluster has been found, the next one can be sought in the remainder of the set I, to obtain a cluster partition – this amounts to the iterated application of Anomalous pattern in Intelligent K-Means (section 3.3). Another version of the one-by-one Anomalous clustering, with the residual data matrix and not necessarily non-overlapping clusters, is described in [90], section 4.6.

5.5.4 Anomalous pattern versus Splitting

A question related to the Anomalous pattern method is whether any relation exists between its criterion (5.54) and Ward criterion λ_w, in its form (4.6) used in the Splitting by separating algorithm. Both methods separate a cluster from a body of entities with a square-error criterion.

To analyze the situation at any of the major iterations, let us denote the entity set under consideration by J, the separated cluster by S_1 with its centroid c_1. Consider also a pre-specified point c in J, which is the centroid of J in Ward clustering and a reference point in the Anomalous pattern clustering. The other subset, $J - S_1$, will be denoted by S_2. In these denotations, the Anomalous pattern criterion can be expressed as:

$$W = \sum_{i \in S_1} d(y_i, c_1) + \sum_{i \in S_2} d(y_i, c)$$

and the Ward splitting criterion as:

$$\lambda_w = \frac{N_J N_1}{N_2} d(c_1, c)$$

where N_J, N_1 and N_2 are the sizes (cardinalities) of J, S_1 and S_2, respectively.

To analyze the relationship between these criteria, let us add to and subtract from W the complementary part $\sum_{i \in S_1} d(y_i, c)$. Then $W = T(Y, c) + \sum_{i \in S_1} d(y_i, c_1) - \sum_{i \in S_1} d(y_i, c)$ where $T(Y, c)$ denotes the sum of squared Euclidean distances from all points in J to c, that is, the scatter of the data around c, an item which is constant with respect to the cluster S_1 being separated. By noting that $d(y_i, c) = (y_i, y_i) + (c, c) - 2(y_i, c)$ and $d(y_i, c_1) = (y_i, y_i) + (c_1, c_1) - 2(y_i, c_1)$, the last expression can be equivalently rewritten as $W = T(Y, c) - N_1 d(c_1, c)$. The following is proven:

Statement 5.17. *Both separating methods, Anomalous pattern and Splitting by separation, maximize the weighted distance $\alpha d(c_1, c)$ between centroid c_1 of the cluster being separated and a pre-specified point c. The difference between the methods is that:*

(1) the weight is $\alpha = N_1$ in the Anomalous pattern and $\alpha = N_1/N_2$ in the Splitting by separation, and

(2) c is the user-specified reference point in the former and the unvaried centroid of set J being split in the latter.

The difference (2) disappears if the reference point has been set at the centroid of J. The difference (1) is fundamental: as proven in section 5.4.1, both sets, S_1 and S_2, are tight clusters in Splitting, whereas only one of them, S_1, is to be tight as the Anomalous pattern: the rest, $J - S_1$, may be a set of rather disconnected entities. This shows once again the effect of size coefficients in cluster criteria.

5.5.5 One-by-one clusters for similarity data

Let $A = (a_{ij})$, $i, j \in I$, be a given similarity matrix and $\lambda\mathbf{z}=(\lambda z_i z_j)$ a weighted binary matrix defined by a binary membership vector $z = (z_i)$ of a subset $S = \{i \in I : z_i = 1\}$ along with its numeric intensity weight λ. When A can be considered as noisy information on a set of weighted "additive clusters" $\lambda_k z_k z_k^T$, $k = 1, ...K$, the following model can be assumed ([121], [88]):

$$a_{ij} = \lambda_1 z_{i1} z_{j1} + ... + \lambda_m z_{iK} z_{jK} + e_{ij} \tag{5.55}$$

where e_{ij} are residuals to be minimized. According to model (5.55), similarities a_{ij} are defined by the intensity weights of clusters containing both i and j. Equations (5.55) must hold, generally, for each pair (i, j). With no loss of generality, we accept in this section that a_{ij} are symmetric and not defined at $i = j$ so that only $i < j$ are considered.

In matrix terms, the model is $A = Z\Lambda Z^T + E$ where Z is an $N \times K$ matrix of cluster membership vectors, Λ is a $K \times K$ diagonal matrix with λ_1,..., λ_K on its main diagonal, all other entries being zero, and $E = (e_{ij})$ the residual matrix. In fact, this matrix equation is analogous to the so-called spectral, or latent value, decomposition of a symmetric matrix A [38].

The one-by-one PCA strategy suggests that, provided that A has been preliminarily centered, minimizing $L(E) = \sum_{i,j} e_{ij}^2$ over unknown S_k and λ_k, $k = 1, ..., K$, can be done by finding one cluster S at a time to minimize

$L = \sum_{i,j \in I}(a_{ij} - \lambda z_i z_j)^2$ with respect to all possible λ and binary $z = (z_i)$. Obviously, given $S \subseteq I$, λ minimizing criterion L is the average similarity within cluster S, that is, $\lambda = a(S)$ where $a(S) = \sum_{i,j \in S; i<j} a_{ij}/[|S|(|S|-1)/2]$. In matrix terms, $a(S) = z^T Az/z^T z$.

After a cluster S and corresponding $\lambda = a(S)$ are found, similarities are changed for residual similarities $a_{ij} - \lambda z_i z_j$, and the process of finding a cluster S and its intensity $a(S)$ is reiterated using the updated similarity matrix. In the case of non-overlapping clusters, there is no need to calculate the residual similarity: the process needs only be applied to only those entities that have remained as yet unclustered.

When the clusters are assumed to be mutually nonoverlapping (that is, the membership vectors z_t are mutually orthogonal) or when fitting of the model is done using the one-by-one PCA strategy, the data scatter decomposition holds as follows [88]:

$$T(A) = \sum_{k=1}^{K} [z_k^T A_k z_k / z_k^T z_k]^2 + L(E) \tag{5.56}$$

in which the least-squares optimal λ_k's are taken as the within cluster averages of the (residual) similarities. The matrix A_k in (5.56) is either A unchanged, if clusters are not overlapping, or a residual similarity matrix at iteration k, $A_k = A - \sum_{t=1}^{k-1} \lambda_k z_k z_k^T$.

Equation (5.56) shows that finding least-squares optimal clusters requires maximizing the intermediate term in (5.56), which is the sum of squared criteria (5.36) or (5.34) discussed in section 5.4.1. When applying the one-by-one extraction strategy so that one cluster is sought at a time, the constituent one cluster criterion is the squared criterion (5.34). To find a cluster optimizing it, one may exploit its square root, criterion (5.34) itself, which will be briefly discussed next.

In [88, 90] a number of incremental procedures to maximize $A(S)$ (5.34) or its square are described to involve the average similarity of an entity $i \in I$ to a subset S,

$$a(i,S) = \sum_{j \in S} a_{ij}/|S| \tag{5.57}$$

A very straightforward procedure for extracting a cluster, given a similarity matrix $A = (a_{ij})$, $i, j \in I$ is referred to as ADDI-S in [88]. The computation starts at S consisting of just two entities, i^* and j^*, such that $a_{i^*j^*}$ is the maximum similarity in A. Given a current S, the procedure proceeds in calculating $a(i,S)$ for all $i \in I$, both belonging to S and not. This can be done incrementally. Then a threshold $\pi = a(S)/2$ is calculated to select those i's that satisfy $a(i,S) > \pi$; this, basically, checks the signs of the attractions of entities to S.

Recall that the attraction of i to S is defined as $\beta(i,S) = a(i,S) - a(S)/2$. Statement 5.10. states that if subset S maximizes criterion (5.34) then entities with $\beta(i,S) > 0$ belong in S and those with $\beta(i,S) < 0$ do not. This line of action can be wrapped up in the following algorithm.

ADDI-S or **Additive similarity cluster**
1. *Initialization.* Find maximum a_{ij} in A and set S to consist of the row and column indexes i, j involved; define $a(S)$ as the maximum a_{ij}. Calculate $a(i,S)$ for all $i \in I$.
2. *Direction of the steepest ascent.* Find $i^* \in I$ that maximizes $\Delta_{i^*} = s_{i^*}\beta(i^*,S) = s_{i^*}(a(i^*,S) - a(S)/2)$ where $s_{i^*} = -1$ if $i \in S$ and $s_{i^*} = 1$, otherwise. If $\Delta_{i^*} \leq 0$, stop the process at current S; otherwise, proceed.
3. *Steepest ascent.* Update S by adding i^* to S if it does not belong to S or by removing it from S if it does. Recalculate $a(S)$ and $a(i,S)$ for all $i \in I$ and return to Step 2.

What is nice in this algorithm is that it involves no ad hoc parameters. Obviously, at the resulting cluster, attractions $\beta(i,S) = a(i,S) - a(S)/2$ are positive for within-cluster elements i and negative for out-of-cluster elements i. Actually, the procedure is a local search algorithm for maximizing criterion $A(S)$ (5.34). This method works well in application to real world data; similar procedures have been successfully applied in [141].

5.6 Overall assessment

An outline of a theory for data visualization and clustering techniques is presented. The theory embraces such seemingly unrelated methods as Principal component analysis, K-Means clustering, hierarchical clustering, conceptual clustering, etc. All these appear to be methods for fitting the same bilinear model of data with different specifications and local search strategies.

The clustering model involves independent representation of the data entries via cluster centroids with residuals, each of which should be made as small as possible. This multi-criterion goal is then reduced to a one-criterion formulation by combining the residuals into the least squares criterion for the follow-up minimization. It is this criterion which determines described rules for data pre-processing, cluster finding and interpretation. The choice of the scalarized criterion is somewhat substantiated by the properties of solutions, which include: averaging quantitative features and taking proportions of categorical features in centroids, using the squared Euclidean distance as the dissimilarity and inner product as the similarity, producing contributions of features to partitions equal to the correlation ratio or Pearson chi-square coefficient, etc.

The theory supports:

1. A unified framework for K-Means and Ward clustering that not only justifies conventional partitioning and agglomerative methods but extends them to mixed scale and multiple table data;
2. Modified versions of the algorithms such as scalable divisive clustering as well as intelligent versions of K-Means mitigating the need in user-driven ad hoc parameters;
3. Compatible measures and criteria for similarity data to produce provably tight clusters with the tightness measured by the attraction coefficient derived in the data recovery framework;
4. Effective measures and clustering criteria for analyzing binary data tables;
5. One-by-one clustering procedures that allow for more flexible clustering structures including single clusters, overlapping clusters, and incomplete clustering;
6. Interpretation aids based on decompositions of the data scatter over elements of cluster structures to judge elements' relative contributions;
7. Conventional and modified measures of statistical association, such as the correlation ratio and Pearson chi-square contingency coefficient, as summary contributions of cluster structures to the data scatter;
8. A related framework for cluster analysis of contingency and flow data in which the data scatter and contributions are measured in terms of the Pearson chi-square contingency coefficient.

Chapter 6

Different Clustering Approaches

After reading through this chapter the reader will know about the most popular other clustering approaches including:

1. Partitioning around medoids (PAM)
2. Gaussian mixtures and the Expectation-Maximization (EM) method
3. Kohonen's Self-organizing map (SOM)
4. Fuzzy clustering
5. Regression-wise K-Means
6. Single linkage clustering and Minimum Spanning Tree (MST)
7. Core and shelled core clusters
8. Conceptual description of clusters

Base words

Association rule A production rule $A \to B$ for a set of, for example, warehouse transactions in which A and B are nonoverlapping sets of goods. It is characterized by its support (proportion of transactions containing A in the set of all transactions) and precision (proportion of transactions containing B among all those containing A): the greater the support and precision, the better the quality of the association rule.

Comprehensive description A conjunctive predicate A describing a subset $S \subset I$ in such a way that A holds for an entity if and only if it belongs to S. It may have errors of two types: the false positives, the entities satisfying A but not belonging to S, and the false negatives, the entities from S that do not satisfy A.

Conceptual description Description of clusters with feature based logical predicates.

Connected component Part of a graph or network comprising a maximal subset of vertices that can be reached from each other by using the graph's edges only.

Decision tree A highly popular concept in machine learning and data mining. A decision tree is a conceptual description of a subset $S \subset I$ or a partition on I in a tree-like manner, the root of the tree corresponding to all entities in I, with each node divided according to values of a feature so that tree leaves correspond to individual classes.

Expectation and maximization EM An algorithm of alternating maximization applied to the likelihood function for a mixture of distributions model. At each iteration, EM is performed according to the following steps: (1) Expectation: Given parameters of the mixture p_k and individual density functions a_k, find posterior probabilities for observations to belong to individual clusters, g_{ik} ($i \in I, k = 1, ..., K$); (2) Maximization: given posterior probabilities g_{ik}, find parameters p_k, a_k maximizing the likelihood function.

Fuzzy clustering A clustering model at which entities $i \in I$ are assumed to belong to different clusters k with a degree of membership z_{ik} so that (1) $z_{ik} \geq 0$ and $\sum_{k=1}^{K} z_{ik} = 1$ for all $i \in I$ and $k = 1, ..., K$. Conventional 'crisp' clusters can be considered a special case of fuzzy clusters in which z_{ik} may only be 0 or 1.

Gaussian distribution A popular probabilistic cluster model characterized by the modal point of the cluster mean vector and the feature-to-feature covariance matrix. The surfaces of equal density for a Gaussian distribution are ellipses.

Graph A formalization of the concept of network involving two sets: a set of nodes, or vertices, and a set of edges or arcs connecting pairs of nodes.

Medoid An entity in a cluster with the minimal summary distance to the other within cluster elements. It is used as a formalization of the concept of prototype in the PAM clustering method.

Minimum spanning tree A tree structure within a graph such that its summary edge weights are minimal. This concept is the backbone of the method of single link clustering.

Mixture of distributions A probabilistic clustering model according to which each cluster can be represented by a unimodal probabilistic density function so that the population density is a probabilistic mixture of the individual cluster functions. This concept integrates a number of simpler models and algorithms for fitting them, including the K-Means clustering algorithm. The assumptions leading to K-Means can be overly restrictive, such as the compulsory z-score data standardization.

Production rule A predicate $A \to B$ which is true for those and only those entities that either satisfy B whenever they satisfy A. A conceptual description of a cluster S can be a production rule $A \to B$ in which B expresses belongingness of an entity to S.

Regression-wise clustering A clustering model in which a cluster is sought as a set of observations satisfying a regression function. Such a model can be fitted with an extension of K-Means in which prototypes are presented as regression functions.

Self-organizing map A model for data visualization in the form of a grid on plane in which entities are represented by grid nodes reflecting both their similarity and the grid topology.

Shelled core A clustering model according to which the data are organized as a dense core surrounded by shells of a lesser density.

Single link clustering A method in clustering in which the between-cluster distance is defined by the nearest entities. The method is related to the minimum spanning trees and connected components in the corresponding graphs.

Weakest link partitioning A method for sequentially removing "weakest link" entities from the data set. Thus found series is utilized for finding the shelled core.

6.1 Extensions of K-Means clustering

6.1.1 Clustering criteria and implementation

Traditionally, a cluster is understood as a subset of entities which are similar to each other and dissimilar from the rest. However, in real-world situations additional properties may be required which may complement and sometimes even contradict this view. Consider, for instance, a set of young men with only one variable, their height, measured. Depending on our perspective, we may be interested in the deviate clusters of those who are too short or too tall to carry out regular sailor duties, or rather in the central core cluster of those who normally fit them. The issue of definition of what is a central core cluster becomes less trivial when encountered in the setting of a web-based or other network. Defining what is a deviate cluster can be of issue in specific settings such as dealing with banking customers. Extending the men's height example to a mixed sex group, one might be interested in defining clusters of male and female individuals in terms of probabilistic normal distributions with differing means and variances, which is a traditional abstraction for height measure on humans. A different clustering goal invites a different formalization and different methods for clustering.

In Chapter 1, a number of user-defined goals for clustering were described. These goals typically cannot be explicitly formalized and, thus, are implicit in clustering methods. The explicit goals are much more technical and down to earth. In particular, criteria optimized by K-Means and Ward methods are such that the found clusters:

1. consist of entities that are the most similar to each other and their centroids;

2. are the most associated with features involved in the data;

3. are the most anomalous and thus interesting.

These claims can be supported with materials in sections 3.1.2, 3.4.4 and 3.3, respectively. A review of related clustering methods emerging in the context of data compression is given in [108].

The goals above obviously could be pursued with differently defined concepts of distance (dissimilarity), centroid and correlation. Even within the data recovery approach, different approximation criteria would lead to differently defined distances and centroids within the goal (1) of the list above. If, for example, the quality of the data recovery is scored with the sum of moduli, rather than squares, of differences between the original and recovered data, then the corresponding distance between entities $x = (x_1, ..., x_m)$ and $y = (y_1, ..., y_M)$ is measured by the sum of moduli, $d = \sum_v |x_v - y_v|$, not the sum of squares.

This distance is frequently used in data analysis and referred to as Manhattan distance or city-block metric. Within the data recovery approach with the least-moduli criterion the concept of centroid would slightly change too: it is the medians, not the averages that would populate them! Both Manhattan distance and median-based centroids bring more stability to cluster solutions, especially with respect to outliers. However, the very same stability properties can make least-moduli based data-recovery clustering less versatile with respect to the presence of mixed scale data [91].

One may use different distance and centroid concepts with computational schemes of K-Means and hierarchical clustering without any strictly defined model framework as, for instance, proposed in [62]. A popular method from that book, PAM (Partitioning around medoids), will be described in the next subsection.

There can be other clustering goals as well. In particular, the following are rather popular:

I Cluster membership of an entity may not necessarily be confined to one cluster only but shared among several clusters (Fuzzy clustering);

II Geometrically, cluster shapes may be not necessarily spherical as in the classic K-Means but may have shapes elongated along regression lines (Regression-wise clustering);

III Data may come from a probabilistic distribution which is a mixture of unimodal distributions that are to be separated to represent different clusters (Expectation–Maximization method EM);

IV A visual representation of clusters on a two-dimensional screen can be explicitly embedded in clustering (Self-Organizing Map method SOM).

Further on in this section we present extensions of K-Means clustering techniques that are oriented towards these goals.

6.1.2 Partitioning around medoids PAM

K-Means centroids are average points rather than individual entities, which may be considered too artificial in some clustering problems in which the user may wish to involve nothing artificial but only genuinely occurring entities. To implement this idea, let us change the concept of a cluster's centroid for that of a cluster's medoid. An entity of a cluster S, $i^* \in S$, will be referred to as its medoid if it minimizes the sum of distances to other elements of S, that is, $\sum_{j\in S} d(i^*,j) = \min_{i\in S} \sum_{j\in S} d(i,j)$. The symbol $d(i,j)$ is used here to denote a dissimilarity function, not necessarily squared Euclidean distance, between entities $i, j \in I$.

Having this concept in mind, the method of partitioning around medoids PAM from [62] can be formulated analogously to that of Straight K-Means. Our formulation slightly differs from the formulation in [62], though it is equivalent to the original, to make its resemblance to K-Means more visible.

Partitioning around medoids PAM
1. *Initial setting.* Choose the number of clusters, K, and select K entities $c_1, c_2, ..., c_K \in I$ with a special algorithm Build. Assume initial cluster lists S_k are empty.
2. *Clusters update.* Given K medoids $c_k \in I$, determine clusters S'_k $(k = 1, ..., K)$ with the Minimum distance rule applied to dissimilarity $d(i, j), i, j \in I$.
3. *Stop-condition.* Check whether $S' = S$. If yes, end with clustering $S = \{S_k\}$, $c = (c_k)$. Otherwise, change S for S'.
4. *Medoids update.* Given clusters S_k, determine their medoids c_k $(k = 1, ..., K)$ and go to Step 2.

The Build algorithm [62] for selecting initial seeds proceeds in a manner resembling that of the iterated Anomalous pattern. It starts with choosing an analogue to the grand mean, that is, the medoid of set I, and takes it as the first medoid c_1. Let us describe how c_{m+1} is selected when a set C_m of m initial medoids, $C_m = \{c_1, ..., c_m\}$, have been selected already ($1 \leq m < K$). For each $i \in I - C_m$, a cluster S_i is defined to consist of entities j that are closer to i than to C_m. The distance from j to C_m is taken as $D(j, C_m) = \min_{k=1}^{m} d(j, c_k)$, and S_i is defined as the set of all $j \in I - C_m$ for which $E_{ji} = D(j, C_m) - d(i, j) > 0$. The summary value $E_i = \sum_{j \in S_i} E_{ji}$ is used as a decisive characteristic of remoteness of i from C_m. The next seed c_{m+1} is defined as the most remote from C_m, that is, an entity i for which E_i is maximum over $i \in I - C_m$.

There is a certain similarity between selecting initial centroids in iK-Means and initial medoids with Build. But there are certain differences as well:

1. K must be pre-specified in Build and not necessarily in iterated Anomalous clusters in iK-Means;

2. The central point of the entire set I is taken as an initial seed in Build and is not in iK-Means;

3. Adding a seed is based on different criteria in the two methods.

Example 6.46. Partitioning around medoids for Masterpieces

Let us apply PAM to the matrix of entity-to-entity distances for Masterpieces displayed in Table 6.1, which is replicated from Table 3.9. We start with building three initial medoids. First, we determine that HF is the medoid of the entire set I, because its total distance to the others, 9.60, is the minimum of total distances in the bottom line of Table 6.1. Thus, HF is the first initial seed.

Table 6.1: Distances between Masterpieces data from Table 3.2.

	OT	DS	GE	TS	HF	YA	WP	AK
OT	0.00	0.51	0.88	1.15	2.20	2.25	2.30	3.01
DS	0.51	0.00	0.77	1.55	1.82	2.99	1.90	2.41
GE	0.88	0.77	0.00	1.94	1.16	1.84	1.81	2.38
TS	1.15	1.55	1.94	0.00	0.97	0.87	1.22	2.46
HF	2.20	1.82	1.16	0.97	0.00	0.75	0.83	1.87
YA	2.25	2.99	1.84	0.87	0.75	0.00	1.68	3.43
WP	2.30	1.90	1.81	1.22	0.83	1.68	0.00	0.61
AK	3.01	2.41	2.38	2.46	1.87	3.43	0.61	0.00
Total	12.30	12.95	10.78	10.16	9.60	13.81	11.35	16.17

Then we build clusters S_i around all other entities $i \in I$. To build S_{OT}, we take the distance between OT and HF, 2.20, and see that entities DS, GE, and TS have their distances to OT smaller than that, which makes them OT's cluster with $E_{OT} = 4.06$. Similarly, S_{DS} is set to consist of the same entities, but its summary remoteness $E_{DS} = 2.98$ is smaller. Cluster S_{GE} consists of OT and DS with even smaller $E_{GE} = 0.67$. Cluster S_{YA} is empty and those of WP and AK contain just another Tolstoy's novel each contributing less than E_{OT}. This makes OT the next selected seed.

After the set of seeds has been updated by OT, we start building clusters S_i again, on the remaining six entities. Of them, clusters S_{DS} and S_{YA} are empty and the others are singletons of which S_{AK} consisting of WP is the most remote, $E_{AK} = 1.87 - 0.61 = 1.26$. This completes the set of initial seeds: HF, OT, and AK. Note, these are novels by different authors.

With the selected seeds, the Minimum distance rule produces the author-based clusters (Step 2). The Stop-condition sends us to Step 4, because these clusters differ from the initial, empty, clusters. At Step 4, clusters' medoids are selected: they are obviously DS in the Dickens cluster, YA in the Mark Twain cluster and either AK or WP in the Tolstoy cluster. With the set of medoids changed to DS, YA and AK, we proceed to Step 2, and again apply the Minimum distance rule, which again leads us to the author-based clusters. This time the Stop-condition at Step 3 halts the process. □

6.1.3 Fuzzy clustering

A fuzzy cluster is represented by its membership function $z = (z_i)$, $i \in I$, in which z_i ($0 \leq z_i \leq 1$) is interpreted as the degree of membership of entity i to the cluster. This extends the concept of a usual, hard (crisp) cluster, which can be considered a special case of the fuzzy cluster corresponding to membership z_i restricted to only 1 or 0 values. A set of fuzzy clusters $z_k = (z_{ik})$, $k = 1, ..., K$, forms a fuzzy partition if the total degree of membership for any entity $i \in I$ is 1, $\sum_k z_{ik} = 1$. This requirement follows the concept of a crisp partition in which any entity belongs to one and only one cluster so that for every i $z_{ik} = 0$ for all k but one. In a fuzzy partition, the full degree of membership of an entity is also 1, but it may be distributed among different clusters. This concept is especially easy to grasp if membership z_{ik} is considered as the probability of

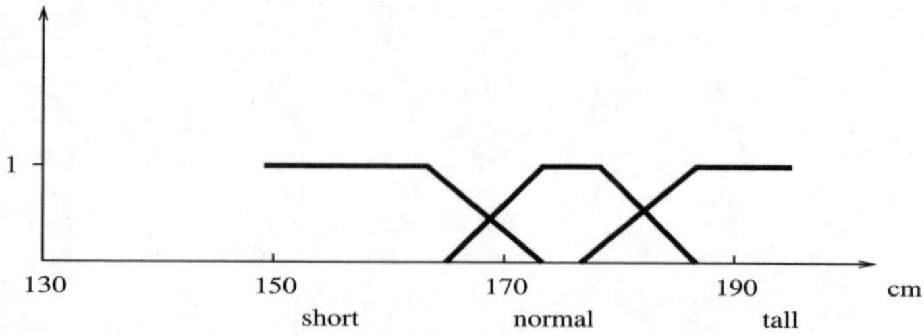

Figure 6.1: Fuzzy sets: Membership functions for the concepts of short, normal and tall in men's height.

belongingness. However, in many cases fuzzy partitions have nothing to do with probabilities. For instance, dividing all people by their height may involve fuzzy categories "short," "average" and "tall" with fuzzy meanings such as shown in Figure 6.1. Fuzzy clustering can be of interest in applications related with natural fuzziness of the cluster borders such as image analysis, robot planning, geography, etc.

The following fuzzy version of Straight K-Means became popular. It involves a fuzzy K-class partition and cluster centroids. The fuzzy partition is represented with an $N \times K$ membership matrix (z_{ik}) $(i \in I, k = 1, ...K)$ where z_{ik} is the degree of membership of entity i in cluster k satisfying conditions: $0 \leq z_{ik} \leq 1$ and $\sum_{k=1}^{K} z_{ik} = 1$ for every $i \in I$. With these conditions, one may think of the total membership of item i as a unity that can be differently distributed among centroids.

The criterion of quality of fuzzy clustering is a modified version of the square-error criterion (3.2),

$$W_F(z, c) = \sum_{k=1}^{K} \sum_{i \in I} z_{ik}^{\alpha} d(y_i, c_k) \tag{6.1}$$

where $\alpha > 1$ is a parameter affecting the shape of the membership function and d distance (2.17) (Euclidean distance squared); as usual, y_i is an entity point and c_k a centroid. In computations, typically, the value of α is put at 2.

By analogy with Straight K-Means, which is an alternating optimization technique, Fuzzy K-Means can be defined as the alternating minimization technique for function (6.1). The centroids are actually weighted averages of the entity points, while memberships are related to the distances between entities and centroids. More precisely, given centroids, $c_t = (c_{tv})$, the optimal membership values are determined as

$$z_{it} = 1 / \sum_{t'=1}^{K} [d(y_i, c_t)/d(y_i, c'_t)]^{\frac{2}{\alpha-1}} \tag{6.2}$$

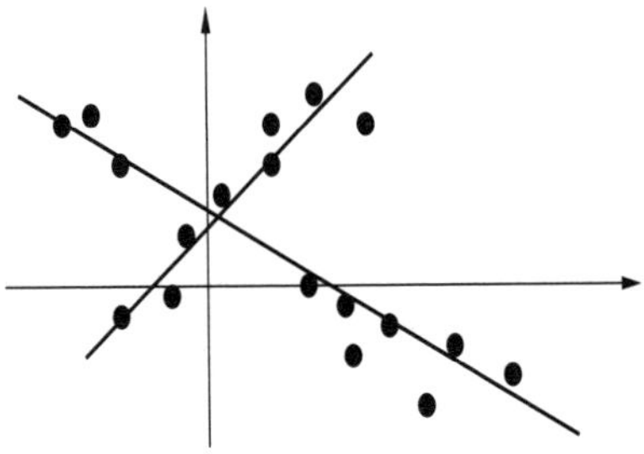

Figure 6.2: Regression-wise clusters: solid lines as centroids.

Given membership values, centroids are determined as convex combinations of the entity points:

$$c_t = \sum_{i \in I} \lambda_{it} y_i \tag{6.3}$$

where λ_{it} is a convex combination coefficient defined as $\lambda_{it} = z_{it}^{\alpha} / \sum_{i' \in I} z_{i't}^{\alpha}$. These formulas follow from the first-degree optimality conditions for criterion (6.1).

Thus, starting from a set of initial centroids and repeatedly applying formulas (6.2) and (6.3), a computational algorithm has been proven to converge to a local optimum of criterion (6.1).

Further improvements of the approach are reported in [60] and [35]. Criterion (6.1) as it stands cannot be associated with a data recovery model. An attempt to build a criterion fitting into the data recovery approach is made in [104].

6.1.4 Regression-wise clustering

In general, centroids c_k can be defined in a space which is different from that of the entity points y_i. Such is the case of regression-wise clustering. Let us recall that a regression function $x_n = f(x_1, ..., x_{n-1})$ may relate a target feature, x_n, to (some of the) other features $x_1, ..., x_{n-1}$ as, for instance, the price of a product to its consumer value and production cost attributes. In regression-wise clustering, entities are grouped together according to the degree of their correspondence to a regression function rather than according to their closeness to the gravity center. That means that regression functions play the role of centroids in regression-wise clustering.

Let us consider a version of Straight K-Means for regression-wise clustering to involve linear regression functions relating standardized y_n to other variables, $y_1, ..., y_{n-1}$, in each cluster. Such a function is defined by the equation $y_n = a_1y_1 + a_2y_2 + ... + a_{n-1}y_{n-1} + a_0$ for some coefficients $a_0, a_1, ..., a_{n-1}$. These coefficients form a vector, $a = (a_0, a_1, ..., a_{n-1})$, which can be referred to as the regression-wise centroid. When a regression-wise centroid is given, its distance to an entity point $y_i = (y_{i1}, ..., y_{in})$ is defined as $r(i, a) = (y_{in} - a_1y_{i1} - a_2y_{i2} - ... - a_{n-1}y_{i,n-1} - a_0)^2$, the squared difference between the observed value of y_n and that calculated from the regression equation. To determine the regression-wise centroid $a(S)$, given a cluster list $S \subset I$, the standard technique of multivariate linear regression analysis is applied, which is but minimizing the within cluster summary residual $\sum_{i \in S} r(i, a)$ over all possible a.

Then Straight K-Means can be applied with the only changes being that: (1) centroids must be regression-wise centroids and (2) the entity-to-centroid distance must be $r(i, a)$.

6.1.5 Mixture of distributions and EM algorithm

Data of financial transactions or astronomic observations can be considered as a random sample from a (potentially) infinite population. In such cases, the data structure can be analyzed with probabilistic approaches of which arguably the most radical is the mixture of distributions approach.

According to this approach, each of the yet unknown clusters k is modeled by a density function $f(x, a_k)$ which represents a family of density functions over x defined up to a parameter vector a_k. A one-dimensional density function $f(x)$, for any small $dx > 0$, assigns probability $f(x)dx$ to the interval between x and $x + dx$; multidimensional density functions have similar interpretation.

Usually, the density $f(x, a_k)$ is considered unimodal (the mode corresponding to a cluster standard point), such as the normal, or Gaussian, density function defined by its mean vector μ_k and covariance matrix Σ_k:

$$f(x, a_k) = (2^p \pi^p |\Sigma_k|)^{-1/2} \exp\{-(x - \mu_k)^T \Sigma_k^{-1} (x - \mu_k)/2\} \qquad (6.4)$$

The shape of Gaussian clusters is ellipsoidal because any surface at which $f(x, a_k)$ (6.4) is constant satisfies equation $(x - \mu_k)^T \Sigma_k^{-1} (x - \mu_k) = const$ defining an ellipsoid. The mean vector μ_k specifies the k-th cluster's location.

The mixture of distributions clustering model can be set as follows. The row points $y_1, ..., y_N$ are considered a random sample of $|V|$-dimensional observations from a population with density function $f(x)$ which is a mixture of individual cluster density functions $f(x, a_k)$ $(k = 1, ..., K)$ so that $f(x) = \sum_{k=1}^K p_k f(x, a_k)$ where $p_k \geq 0$ are the mixture probabilities, $\sum_k p_k = 1$. For $f(x, a_k)$ being the normal density, $a_k = (\mu_k, \Sigma_k)$ where μ_k is the mean and Σ_k the covariance matrix.

To estimate the individual cluster parameters, the main approach of mathematical statistics, the maximum likelihood, is applied. The approach is based on the postulate that really occurred events are those that are most likely. In its simplest version, the approach requires the finding of the parameters $p_k, a_k,\ k = 1, ..., K$, by maximizing the logarithm of the likelihood of the observed data under the assumption that the data come from a mixture of distributions:

$$L = \log\{\prod_{i=1}^{N} \sum_{k=1}^{K} p_k f(y_i, a_k)\}.$$

To computationally handle the maximization problem, this criterion can be reformulated as

$$L = \sum_{i=1}^{N} \sum_{k=1}^{K} g_{ik} \log p_k + \sum_{i=1}^{N} \sum_{k=1}^{K} g_{ik} \log f(y_i, a_k) - \sum_{i=1}^{N} \sum_{k=1}^{K} g_{ik} \log g_{ik} \quad (6.5)$$

where g_{ik} is the posterior density of class k, defined as

$$g_{ik} = \frac{p_k f(y_i, a_k)}{\sum_k p_k f(y_i, a_k)}.$$

In this way, criterion L can be considered a function of two groups of variables: (1) p_k and a_k, and (2) g_{ik}, so that the method of alternating optimization can be applied. The alternating optimization algorithm for this criterion is referred to as the EM-algorithm since computations are performed as a sequence of the so-called estimation (E) and maximization (M) steps.

EM-algorithm
Start: With any initial values of the parameters,
E-step: Given p_k, a_k, estimate g_{ik}.
M-step: Given g_{ik}, find p_k, a_k maximizing the log-likelihood function (6.5).
Halt: When the current parameter values approximately coincide with the previous ones.

If f is the Gaussian density function, then the optimal values of parameters, in M-step, can be found with the following formulas:

$$\mu_k = \sum_{i=1}^{N} g_{ik} y_i / g_k,\ \Sigma_k = \sum_{i=1}^{N} g_{ik} (y_i - \mu_k)(y_i - \mu_k)^K / g_k$$

where $g_k = \sum_{i=1}^{N} g_{ik}$.

If the user needs an assignment of the observations to the classes, the posterior probabilities g_{ik} can be utilized: i is assigned to that k for which g_{ik} is the maximum. Also, ratios g_{ik}/g_k can be considered as fuzzy membership values.

The situation, in which all covariance matrices Σ_k are diagonal and have the same variance value σ^2 on the diagonal, corresponds to the assumption that all clusters have uniformly spherical distributions. This situation is of particular interest because the maximum likelihood criterion is here equivalent to $W(S, c)$ criterion of K-Means and, moreover, there is a certain homology between the EM and Straight K-Means algorithms.

Indeed, under the assumption that feature vectors corresponding to entities $x_1, ..., x_N$ are randomly and independently sampled from the population, with unknown assignment of the entities to clusters S_k, the likelihood function in this case has the following formula:

$$L(\{\mu_k, \sigma^2, S_k\}) = A \prod_{k=1}^{K} \prod_{i \in S_k} \sigma^{-M} \exp\{-(x_i - \mu_k)^T \sigma^{-2}(x_i - \mu_k)/2\} \quad (6.6)$$

so that to maximize its logarithm, the following function is to be minimized:

$$l(\{\mu_k, \sigma^2, S_k\}) = \sum_{k=1}^{K} \sum_{i \in S_k} (x_i - \mu_k)^T (x_i - \mu_k)/\sigma^2 \quad (6.7)$$

This function is but a theoretic counterpart to K-Means criterion $W(S, \mu) = \sum_{k=1}^{K} \sum_{i \in S_k} d(y_i, \mu_k)$ applied to vectors y_i obtained from x_i with z-scoring standardization (shifting scales to grand means and normalizing them by standard deviations).

Thus the mixture model can be considered a probabilistic model behind the conventional K-Means method. Moreover, it can handle overlapping clusters of not necessarily spherical shapes (see Figure 6.3). Note however that the K-Means data recovery model assumes no restricting hypotheses on the mechanism of data generation. We also have seen how restricting is the requirement of data standardization by z-scoring, associated with the model. Moreover, there are numerous computational issues related to the need in estimating much larger numbers of parameters in the mixture model.

One of the latest and most successful attempts in application of this approach is described in [141]. The authors note that there is a tradeoff between the complexity of the probabilistic model and the number of clusters: a more complex model may fit to a smaller number of clusters. To select the better model one can choose that one which gives the higher value of the likelihood criterion which can be approximately evaluated by the so called Bayesian Information Criterion (BIC) equal, in this case, to

$$BIC = 2 \log p(X/p_k, a_k) - \nu_k \log N \quad (6.8)$$

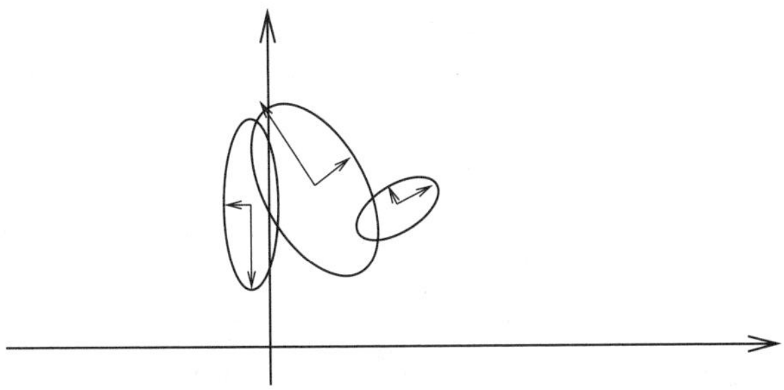

Figure 6.3: A system of ellipsoids corresponding to a mixture of three normal distributions with different means and covariance matrices.

where X is the observed data matrix, ν_k, the number of parameters to be fitted, and N the number of observations, that is, the rows in X. The BIC analysis has been demonstrated to be useful in accessing the number of clusters.

To guarantee normality, the authors applied three popular transformations of the data: logarithm, square root, and z-score standardization of rows (genes), not columns. Typically, with logarithms the data better accord to the Gaussian distribution. The authors note that using complex models cannot be always feasible: the number of parameters estimated per cluster at the space dimension 24 becomes 24+ 24*23/2=324, which can be greater than the cluster's cardinality. Overall the results are inconclusive [141].

Further advances in mixture of distributions clustering are described in [5, 142].

6.1.6 Kohonen self-organizing maps SOM

The Kohonen Self-Organizing Map is an approach to visualize a data cluster structure by mapping it onto a plane grid [67]. Typically, the grid is rectangular and its size is determined by the user-specified numbers of its rows and columns, r and c, respectively, so that there are $r \times c$ nodes on the grid. Each of the grid nodes, e_t $(t = 1, ..., rc)$, is one-to-one associated with the so-called model, or reference, vector m_t which is of the same dimension as the entity points y_i, $i \in I$. Initially, vectors m_t are to be set in the data space either randomly or according to an assumption of the data structure such as, for instance, K-Means centroids. Given vectors m_t, entity points y_i are partitioned according to a version of the Minimum distance rule into sets I_t. For each t^*, I_{t^*} consists of those y_i whose distance to m_{t^*} is minimum over all $t = 1, ..., rc$.

Besides, a neighborhood structure is assigned to the grid. In a typical case, the neighborhood of node e_t is set E_t of all nodes on the grid whose path distance from e_t on the grid is smaller than a pre-selected threshold value.

Historically, all SOM algorithms worked in an incremental manner as neuron networks, but later on, after some theoretical investigation, straight versions appeared, such as the following.

Straight SOM
1. *Initial setting.* Select r and c for the grid and initialize model vectors m_t $(t = 1, ..., rc)$ in the entity space.
2. *Neighborhood update.* For each grid node e_t, define its grid neighborhood E_t and collect the list I_u of entities most resembling the model m_u for each $e_u \in E_t$.
3. *Seeds update.* For each node e_t, define new m_t as the average of all entities y_i with $i \in I_u$ for some $e_u \in E_t$.
4. *Stop-condition.* Halt if new m_ts are close to the previous ones (or after a pre-specified number of iterations). Otherwise go to 2.

As one can see, the process much resembles that of Straight K-Means, with the model vectors similar to centroids, except for two items:

1. The number of model vectors is large and has nothing to do with the number of clusters, which are determined visually in the end as grid clusters.
2. The averaging goes along the grid, not entity space, neighborhood.

These features provide for less restricted visual mapping of the data structures to the grid. On the other hand, the interpretation of results here remains more of an intuition rather than instruction. A framework relating SOM and EM approaches is proposed in [9].

6.2 Graph-theoretic approaches

In this section we present some approaches that are relevant to networks and other structural data (see, for instance, [3]). As the bottom line, they rely on graph-theoretic properties of data.

6.2.1 Single linkage, minimum spanning tree and connected components

The single linkage, or nearest neighbor, method is based on entity-to-entity dissimilarities like those presented in the Primates data (Table 1.2) or in the distance matrix calculated in section 3.2.2 when computing initial seeds for the Masterpieces data (Table 3.9).

It should be pointed out that the dissimilarities can be computed from the original data table at each step so that there is no need to maintain the dissimilarity matrix as a separate data file. This may save quite a lot of memory, especially when the number of entities is large. For instance, if the size of the original data table is 1000 × 10, it takes only 10,000 numbers to store, whereas the entity-to-entity distances may take up to half a million numbers. There is always a trade-off between memory and computation in distance based approaches that may require some efforts to balance.

The single linkage approach is based on the principle that the dissimilarity between two clusters is defined as the minimum dissimilarity (or, maximum similarity) between entities of one and the other cluster [41]. This can be implemented into the agglomerative distance based algorithm described in section 4.4.1 by leaving it without any change except for the formula (4.11) that calculates distances between the merged cluster and the others and must be substituted by the following:

$$d_{w1\cup w2,w} = \min(d_{w1,w}, d_{w2,w}) \tag{6.9}$$

Formula (6.9) follows the principle of minimum distance, which explains the method's name.

In general, agglomerative processes are rather computationally intensive because the minimum of inter-cluster distances must be found at each merging step. However, for single linkage clustering, there exists a much more effective implementation involving the concept of the minimum spanning tree (MST) of a weighted graph.

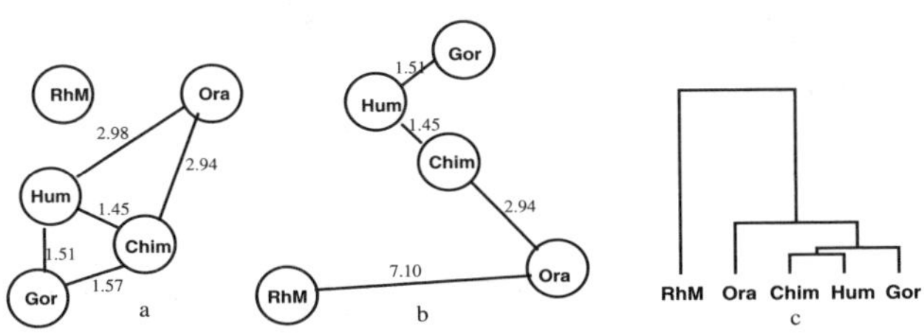

Figure 6.4: Weighted graph (a), minimum spanning tree (b) and single-linkage hierarchy for Primates data (c).

A weighted graph, in our context, is a representation of a dissimilarity or distance matrix $D = (d_{ij})$ with nodes corresponding to entities $i \in I$ and edges connecting any i and j from I with weight d_{ij}. Such a graph for the Primates data is in Figure 6.4 (a): only edges whose weights are smaller than 3 are shown.

A spanning tree of a weighted graph is a subgraph T without cycles such that it covers all its nodes (as in Figure 6.4 (b)). The length of T is defined as the sum of all weights d_{ij} over edges $\{i, j\}$ belonging to T. A minimum spanning tree (MST) T must have minimum length. The concept of MST is of prime importance in many applications in the Computer Sciences. Let us take a look at how it works in clustering.

First of all, let us consider the so-called Prim algorithm for finding an MST. The algorithm processes nodes (entities), one at a time, starting with $T = \emptyset$ and updating T at each step by adding to T an element i (and edge $\{i, j\}$) minimizing d_{ij} over all $i \in I - T$ and $j \in T$. An exact formulation is this.

Prim algorithm
1. *Initialization.* Start with set $T \subset I$ consisting of an arbitary $i \in I$ with no edges.
2. *Tree update.* Find $j \in I - T$ minimizing $d(i, j)$ over all $i \in T$ and $j \in I - T$. Add j and (i, j) with the minimal $d(i, j)$ to T.
3. *Stop-condition.* If $I - T = \emptyset$, halt and output tree T. Otherwise go to 2.

To build a computationally effective procedure for the algorithm may be a cumbersome issue, depending on how $d(i, j)$ and their minima are handled, to which a lot of work has been devoted. A simple pre-processing step can be quite useful: in the beginning, find a nearest neighbor for each of the entities; only they may go to MST.

Example 6.47. Building MST for Primate data

Let us apply Prim algorithm to the Primates distance matrix in Table 1.2, p. 5.

Let us start, for instance, with $T = \{\text{Human}\}$. Among remaining entities Chimpanzee is the closest to Human (distance 1.45), which adds Chimpanzee corresponding edge to T as shown in Figure 6.4 (b). Among the three other entities, Gorilla is the closest to one of the elements of T (Human, distance 1.57). This adds Gorilla to T and the corresponding edge in MST in Figure 6.4 (b). Then comes Orangutan as the closest to Chimpanzee in T. The only remaining entity, Monkey, is nearest to Orangutan, as shown on the drawing. □

Curiously, in spite of its quantitative definition, MST depends only on the order of dissimilarities, not their quantitative values.

To see the relation between an MST and single linkage clustering, let us build an MST based distance between nodes (entities). This distance, $T(i, j)$, is defined as the maximum weight of an edge on the only path joining i and j in the tree T. (If another path existed, the two paths would create a cycle through these two nodes, which is impossible because T is a tree.) The definition implies that $d_{ij} \geq T(i, j)$ for all $i, j \in I$ because, otherwise, the tree T can be made shorter by adding the shorter (i, j) and removing the edge at which $T(i, j)$ is reached. Moreover, $T(i, j)$ can be proven to be an ultra-metric (as defined in section 4.4.1). The metric $T(i, j)$ coincides with that implied by the single linkage method.

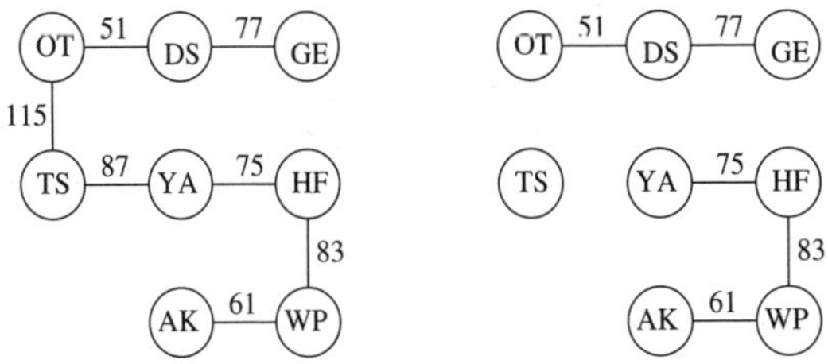

Figure 6.5: Minimum spanning tree for Materpieces (a) and a result of cutting it (b).

An MST T allows finding the single linkage hierarchy by a divisive method: sequentially cutting edges in MST beginning from the largest and continuing in descending order. The result of each cut creates two connected components that are children of the cluster in which the cut occurred (see, for instance, Figure 6.4 (c)).

A peculiarity of the single linkage method is that it involves just $N-1$ dissimilarity entries occurring in an MST rather than all $N(N-1)/2$ of them. This results in a threefold effect: (1) a nice mathematical theory, (2) fast computations, and (3) poor application capability.

Example 6.48. MST and single linkage clusters in the Masterpieces data
To illustrate point (3) above let us consider an MST built on the distances between Masterpieces in Table 3.9 (see Figure 6.5 (a)). Cutting the tree at the longest two edges, we obtain clusters presented in part (b) of the Figure. Obviously these clusters do not reflect the data structure properly, in spite of the fact that the structure of (a) corresponds to authors.

□

One more graph-theoretical interpretation of single-linkage clusters is in terms of a prime cluster concept in graph theory, the connected component. Let us consider a dissimilarity matrix $D = (d_{ij})$, $i, j \in I$. Then, given any real u, the so-called u-threshold graph G_u is defined on the set of vertices I as follows: $i \in I$ and $j \in I$ are connected with edge (i,j) if and only if $d_{ij} < u$. Obviously, the edges in this graph are not directed since $d_{ij} = d_{ji}$. A subset S is referred to as connected in a graph if for any different $i, j \in S$ there exists a path between i and j within S. A connected component is a maximal connected $S \subseteq I$: it is connected and either coincides with I or loses this property if any vertex $i \in I - S$ is added to S.

When a minimum spanning tree T is split over its greatest link leading to a partition of I in two clusters, S_1 and S_2, each of them is obviously a connected component of the threshold graph G_u with u equal to the weight of the link removed from T. For the sake of simplicity, we consider that no other edge weight in T is equal to u. Then any i and j in S_1 can be joined by a path belonging to the relevant part of T, thus G_u, because all the weights along T are smaller than u by definition. On the other hand, for any $k \in S_2$, the dissimilarity $d_{ik} \geq u$ so that edge (i, k) is absent from G_u thus making S_1 a connected component. The fact that S_2 is a connected component of G_u can be proven similarly. When the maximum weight u on the tree T is not unique, T must be split along all maximum weight edges to make the split parts connected components. Then one could prove that the connected components of threshold graphs G_u (for u being the weight of an edge from a minimum spanning tree) are single linkage clusters.

6.2.2 Finding a core

Here we consider the problem of finding a dense core, rather than a deviate pattern, in a given set of interrelated objects. This problem attracted attention not only in data mining but in other disciplines such as Operations Research (knapsack and location problems).

We follow the way to formalize the problem in terms of a, possibly edge-weighted, graph proposed in [102] and further extended in [98]. Such is the graph of feature-to-feature similarities in Figure 6.6. For a subset of vertices $H \subseteq I$ and a vertex $i \in H$, let us define linkage $\pi(i, H)$ as the sum of the weights of all edges connecting i with $j \in H$.

Obviously, the thus defined linkage function $\pi(i, H)$ is monotone over H: adding more vertices to H may only increase the linkage; that is, $\pi(i, H) \leq \pi(i, H \cup G)$ for any vertex subset $G \subseteq I$. All contents of this section are applicable to any monotone linkage function. A monotone linkage function such as $\pi(i, H)$ can be used to estimate the overall density of a subset $H \subseteq I$ by "integrating" its values $\pi(i, H)$ over $i \in H$. In particular, an integral function defined by the weakest link in H,

$$F_\pi(H) = \min_{i \in H} \pi(i, H), \tag{6.10}$$

will be referred to as the tightness function.

Table 6.2: Matrix of column-to-column squared inner products from Table 2.17, multiplied by 100.

	LS	LD	NC	SC	Pe	Ob	Di
LS	54.56	13.21	30.79	8.19	0.03	3.56	4.28
LD	13.21	48.05	21.90	13.90	0.54	12.51	7.85
NC	30.79	21.90	79.01	11.11	14.81	3.70	33.33
SC	8.19	13.90	11.11	351.56	25.52	0.52	18.75
Pe	0.03	0.54	14.81	25.52	39.06	14.06	6.25
Ob	3.56	12.51	3.70	0.52	14.06	39.06	6.25
Di	4.28	7.85	33.33	18.75	6.25	6.25	25.00

Example 6.49. Linkage function on Masterpieces data features

For the pre-processed Masterpieces data in Table 2.17 or 3.2, let us consider the matrix of squared feature-to-feature inner products in Table 6.2.

This matrix can be used for analysis of interrelations between features of Masterpieces. In particular, let us draw a weighted similarity graph whose vertices are features and whose edges correspond to those similarities which are, in the rounded form, 14 or greater, see Figure 6.6. In this graph, at set $H = \{LS, LD, NC, SC\}$, $\pi(LS, H) = 31$, $\pi(LD, H) = 22 + 14 = 36$, $\pi(NC, H) = 31 + 22 = 53$ and $\pi(SC, H) = 14$. The value of tightness function at this set H is the minimum of these four, $F_\pi(H) = 14$, thus making it the core.

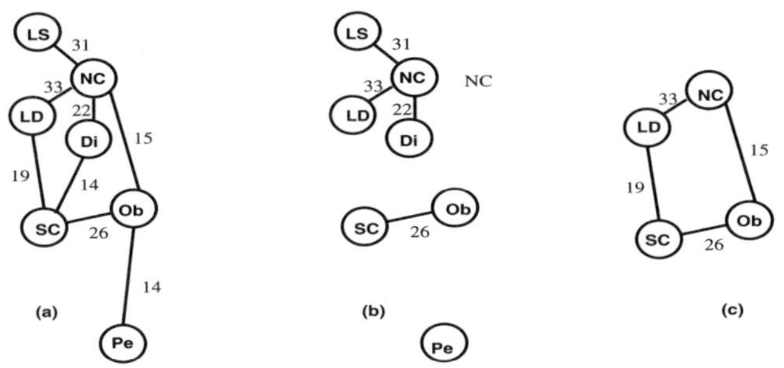

Figure 6.6: Weighted graph generating the summary linkage function π (a); connected components after cutting two weakest edges in its Maximum spanning tree (b); and the core of the tightness function over π (c).

□

A property of the tightness function (easily following from the monotonicity of π) is that it satisfies the so-called quasi-convexity condition: for any $H, G \subseteq I$,

$$F(H \cup G) \geq \min(F(H), F(G)). \tag{6.11}$$

Actually, inequality (6.11) is a characteristic of the class of tightness functions [98].

We define the maximum density core of the graph as H maximizing the tightness function over all $H \subseteq I$. It appears the problem of finding such a set can be solved rather efficiently in a greedy-wise manner. Moreover, not only the maximally dense core of I can be found but, in the same run, also a nested chain of its "shells" with monotonely decreasing densities as well. This "shelled core" can be considered an implementation of the idea of a multi-resolution view at the structure of a system of interrelated elements.

Let us define an F-pattern as a subset $S \subset I$ which is strongly separated from the rest so that its F score decreases if any supplementary elements are added, even if some of its elements are removed, that is, $F(S) > F(H)$ for any $H \subseteq I$ such that $H \cap (I - S) \neq \emptyset$.

Thus defined, F-patterns must be chain-nested.

Statement 6.18. *The set of all F-patterns, P, is nonempty and chain-nested, that is, $S_1 \subseteq S_2$ or $S_2 \subseteq S_1$ for any $S_1, S_2 \in P$.*

Proof: If S_1, S_2 are F-patterns and S_1 is not part of S_2, then $F(S_2) > F(F_1)$. If, moreover, S_2 is not part of S_1, then $F(S_1) > F(S_2)$. The contradiction proves that P must be chain-nested. Besides, $S = I$ makes the definition of F-pattern true because of the false premise, which proves that I is always an F-pattern and completes the proof, q.e.d.

In general, more dense subparts may occur within the smallest F-pattern S: nothing prevents one or more $H \subset S$ with $F(H) > F(S)$ to exist. When this is not the case, that is, when the smallest F-pattern is a global maximizer of the function F, the set of F-patterns will be referred to as a layered cluster, which is uniquely defined and thus can be considered as an aggregate representation of the F density structure in I.

Statement 6.19. *If $F(S)$ is a tightness function, then its minimum pattern is the largest global maximizer of $F(S)$ in the set of all $S \subseteq I$.*

Proof: Let S be the minimum F-pattern in the chain nested set of F-patterns. If S is not a global maximizer of F, then $F(H) > F(S)$ for some $H \subseteq I$. In fact, all such H must fall within S, because of the definition of F-patterns. Let us take a maximal subset $H \subset S$ in the set of all H such that $F(H) > F(S)$ and prove that H is a pattern as well. Indeed, let us take any $S' \subseteq S$ such that $S' \cap (I - H) \neq \emptyset$; the existence of such S' follows from the fact that H does not coincide with S. Then $F(H) > F(H \cup S')$ because of the assumed maximality of H within S. But $F(H \cup S') \geq \min(F(S'), F(H))$ according to (6.11), that is, $F(H) > F(S')$. Let us consider now an S' which is not contained in S and still satisfies the condition $S' \cap (I - H) \neq \emptyset$. (This may only happen when S is not equal to I.) By the definition of S', $F(S) > F(S')$ because S is a pattern. Therefore, $F(H) > F(S')$. This implies that H is a pattern, which contradicts the assumption of minimality of S. Thus, no $H \subset S$ exists with $F(H) > F(S)$ and S is the maximum global maximizer of $F(H)$, q.e.d.

Let us denote by $m(H)$ the "weakest link", that is, the set of elements $i \in H$ at which the value of $F(H)$ is reached:

$$m(H) = \{i : \pi(i, H) = \min_{j \in H} \pi(j, H)\}.$$

Obviously, $m(H)$ is not empty if H is not empty. Iteratively applying the operation m to I, $I - m(I)$, etc., one can build a series of weakest links that will be referred to as the weakest link partition of I.

Weakest Link Partitioning.
Input: Monotone linkage function $\pi(i, H)$ defined for all pairs i, H such that $i \in H \subseteq I$.
Output: Weakest link partition $M = (M_0, M_1, ..., M_n)$ of I along with class values $F(M) = \{F_0, F_1, ..., F_n\}$ defined as follows.
Step 0. *Initial setting.* Put $t = 0$ and define $I_0 = I$.
Step 1. *Partitioning.* Find class $M_t = m(I_t)$ and define $I_{t+1} = I_t - M_t$. Define class value $F_t = F_\pi(I_t) = \pi(i, I_t)$ for $i \in M_t$.
Step 2. *Stop-condition.* If $I_{t+1} = \emptyset$, halt. Otherwise, add 1 to t and go to Step 1.

The layered cluster of F-patterns can be easily extracted from the weakest link partition M thus produced.

From the sequence F, pick up the smallest index t^* among those maximizing F_t, $t = 0, 1, ..., n$. Then apply the same selection rule to the starting part of the sequence F, $F^{t^*} = (F_0, ..., F_{t^*-1})$ obtained by removing F_{t^*} and all the consequent elements. Reiterating this pick-and-removal process until all elements of F are removed, we obtain set T^* of all the picked up indices.

Sets I_{t^*}, $t^* \in T^*$, form the layered cluster of F_π, which is proven in [98].

Example 6.50. Building shelled core over features

Let us apply the Weakest link partitioning algorithm to the graph in Figure 6.6 with linkage function $\pi(i, S) = \sum_{j \in S} a_{ij}$ with a_{ij} being the weight. Obviously, $m(I) = \{Ob\}$ because $\pi(Ob, I) = 14$ is the minimum of $\pi(i, I)$ over all $i \in I$. With the weakest link Ob removed from I, the minimum of $\pi(i, I - \{Ob\})$ is reached at LS with $\pi(LS, I - \{Ob\}) = 31$, which is the next weakest link to be removed. The next entity to be removed is LD, with $\pi(LD, I - \{Ob, LS\}) = 36$. In the remaining set $I_3 = \{NC, Di, SC, Pe\}$, the weakest link is Pe with $\pi(Pe, \{NC, Di, SC, Pe\}) = 41$. This yields $I_4 = \{NC, Di, SC\}$ with the minimum link, 19, reached at $m(I_4) = \{SC\}$. Two remaining entities, NC and Di, are linked by 33. The results can be represented as a labeled sequence:

$$(Ob)^{14}(LS)^{31}(LD)^{36}(Pe)^{41}(SC)^{19}(Di, NC)^{33},$$

where the parentheses contain sets $M_t = m(I_t)$ removed at each step of the algorithm, their order reflecting the order of removals. The labels correspond to the values of the linkage function $F_\pi(I_t)$ for $t = 0, 1, 2, 3, 4, 5$. The maxima are 41, 36, 31, 14; the corresponding patterns being $H_3 = \{NC, Di, SC, Pe\}$ (part (c) on Figure 6.6), $H_2 = H_3 \cup \{LD\}$, $H_1 = H_2 \cup \{LS\}$, and $H_0 = I$, to form the layered cluster.

It should be noted that this drastically differs from any partition along the maximum spanning tree presented in part (b) of Figure 6.6. □

6.3 Conceptual description of clusters

Interpretation of clusters can be done at the following three levels: (1) cluster representative, (2) statistical description, and (3) conceptual description. Of these, we considered (1) and (2) in the previous chapters. Here we concentrate on the conceptual description of clusters.

6.3.1 False positives and negatives

A conceptual description of a cluster is a logic predicate over features defined on entities, that is, in general, true on entities belonging to the cluster and false on entities out of it. For instance, the cluster of Dickens masterpieces, according to Masterpieces data in Table 1.10 can be described conceptually with predicate "SCon=0" or predicate "$19 \leq$ LSent ≤ 30 & $2 \leq$ NChar ≤ 3." These descriptions can be tested for any entity from the table. Obviously, the former predicate distinctively describes the cluster with no errors at all, while the latter admits one false positive error: the entity HuckFinn satisfies the description but belongs to a different cluster. False positive errors are entities that do not belong to the cluster but satisfy its description; in contrast, an entity from the cluster that does not satisfy its description is referred to as a false negative. Thus, the problem of the conceptual description of clusters can be formalized as that of finding as brief and clear descriptions of them as possible while keeping the false positive and false negative errors as low as possible.

Traditionally, the issue of the interpretation of clusters is considered as art rather than science. The problem of finding a cluster description is supposed to have little interest on its own and be of interest only as an intermediate tool in cluster prediction: a cluster description sets a decision rule which is then applied to predict, given an observation, which class it belongs to. This equally applies to both supervised and unsupervised learning, that is, when clusters are pre-specified or found from data. In data mining, the prediction problem is frequently referred to as that of classification so that a decision rule, which is not necessarily based on a conceptual description, is referred to as a classifier (see, for instance, [23]). In clustering, the problem of cluster description is part of the interpretation problem. Some authors even suggest that conceptual descriptions be a form of representing clusters [59].

6.3.2 Conceptually describing a partition

In the literature, the problem of conceptual description of a partition has received by far much more attention than the problem of description of a single cluster, probably because it better fits into the concept of a decision, or classification, tree which is being used as the main device for conceptual description. Once again it should be pointed out that the primary goal for building a decision tree is typically prediction of the partition under consideration rather than its description.

The concept of a classification decision tree (see [120], [23]) is very similar to that of the conceptual clustering tree discussed in section 4.3 except for the splitting criterion. In conceptual clustering, the criterion is to get clusters as homogeneous as possible with regard to all variables. In decision trees, the criterion is homogeneity with regard to a pre-specified quantitative or categorical feature - that one which is to be predicted or interpreted.

The homogeneity is scored with the same types of criteria as in conceptual clustering, e.g., impurity function in CART [11], Pearson chi-squared contingency coefficient in CHAID [42], and entropy in C4.5 [111] (see also [23]). These criteria can be considered differently weighted aggregations of correct and erroneous decisions in the typical situation of a multi-class partition under consideration. To be more precise, let us denote by p_{sk} the proportion of entities that fall in class s of a conceptual description while belonging in fact to class k of the original partition. This decision can be assigned a different weighting coefficient such as $g_{sk} = p(k/s) - p(k)$, $q_{sk} = p(k/s)/p(k) - 1$ and $l_{sk} = \log(p(k/s)/p(k))$. It is not difficult to prove, in the light of results reported in section 5.2.4 that it is exactly these weights averaged according to proportions p_{sk} that correspond to the scoring functions applied in the three decision tree building programs above, CART, CHAID and C4.5, respectively.

A decision tree resulting from learning a pre-specified partition $S = \{S_1, ..., S_k, ..., S_K\}$ is defined in such a way that (a) each of its internal nodes is labeled by a feature, (b) each of its edges is labeled by a predicate pertaining to the feature associated with the parent, and (c) each leaf node is labeled with an assigned class S_k. In this way, each leaf node gets a conceptual description, which is a conjunction of predicates assigned to edges on the path leading from the root to the leaf. Each S_k is conceptually described by descriptions of S_k-labeled leaves combined with the logical operation OR (disjunction) $(k = 1, ..., K)$.

Decision trees are built from top to bottom (the root representing all of the entity set I being considered) in a divisive manner, each time splitting an entity subset, represented by a tree node, according to a feature and its derived predicates. For a categorical feature f, the predicates can express just simple tests such as $f = v$ for each of its categories v, thus dividing the node subset according to categories v. Any algorithm for building a decision tree includes answers to the following questions:

1. Whether any node should be split at all (testing a stopping condition).

2. Which node of the tree and by which variable to split.

3. What class of a cluster partition S is to be assigned to a terminal node?

An answer to question 2 is provided by testing the splitting criterion at each of the nodes with each of the features, still unused at the branch, and selecting the best.

An answer to question 3, on the first glance, can be provided by using the modal (most frequent) cluster. However, this can be sometimes too rough a decision. Assume, for example, a set of individuals whose features are not highly related to each other, such as risk factors with regard to some category S_1 that can be a medical condition such as asthma or a deviant behavior such as theft. In a sample reflecting the population, the proportion of individuals in category S_1 will be extremely low, say, one per cent. Then, in any sizeable subset defined by combining values of different features, the proportion of normal individuals will be always higher than those of category S_1, which would imply that to get a decision tree leaf assigned to S_1, it would have to be a small subset indeed. Most importantly, the S_1 assigned leaves typically will cover an insignificant proportion of category S_1 in the sample. This is why a differential index such as the Quetelet coefficient $q_{1w} = (p(1/w) - p(1))/p(1)$ can be of greater interest in assigning classes of S to leaves w than just the conditional probability $p(1/w)$: the assignment still will be based on the absolute proportion $p(1/w)$ of S_1 in w – in its relation to the average rate $p(1) = |S_1|/N$. (We use 1 rather than S_1 here for the sake of convenience.) The differential assignment index can be selected to match that one used, in the averaged format, for scoring splits of tree clusters at the tree building process.

The question 1. above is the most difficult and it usually gets an ad hoc answer: the splitting process stops when the tree becomes too big and/or leaf classes too small. Then the tree is pruned back by removing splits which lead to the most ambiguous assignments or result in clusters which are considered too small.

The classification tree divides the entity set I in a tree-like manner, each division done by dividing a feature's range into two or more categories. For instance, Figure 6.7 presents a decision tree found by initially partitioning the vertical axis (feature y) in two intervals over point A, and then by partitioning the upper part with b and the lower with c along the horizontal axis (feature x). These divisions are presented in part (a) of the Figure as occurring in the feature space, and in part (b) in a drawn tree. The regions of the space get simple conceptual descriptions so that the description of S defined by this tree, after pruning, is: black circles are in the part "less than A of y," and white circles are in other regions (see part (c) of the Figure). This would lead to four false positives for S and two false negatives for "not S" decisions.

Decision trees are convenient because they provide for easily understandable descriptions. They are also convenient computationally because they rely on statistical characteristics of features, and thus are scalable to large data sizes. When the data size is massive indeed, decision trees still can be built on data subsamples.

There are some shortcomings too. These techniques are 'monothetic' so that each split goes along only one feature, and not directly applicable to clus-

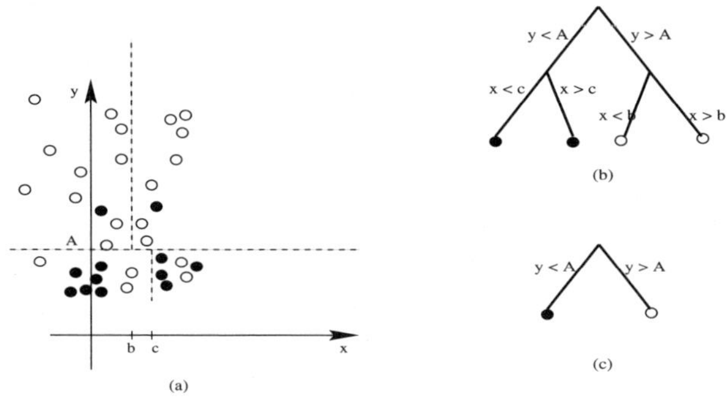

Figure 6.7: Classification tree in the feature space (a); the same, as a decision tree (b); the tree after pruning (c).

ters whose definitions involve combinations of features. Within traditional approaches, handling both continuous-valued and categorical features can be an issue because of incomparability between statistical indexes used for different feature types. The data recovery framework seems a good vehicle for overcoming this hurdle, since different types of features get unified contribution based indexes of association, such as in sections 5.2 and 3.4.4.

One more aspect of decision trees that can be addressed with the data recovery approach is of developing splitting criteria explicitly oriented towards description of a single cluster or a category rather than a partition. The conventional decision tree techniques do not pay much attention to the cases in which the user is interested in getting a description for a cluster S only, though in many applications $I-S$ can be highly non-homogeneous so that its conceptual description may have no meaning at all.

A scoring function of a split of tree cluster S_w in two parts S_1 and S_2, with respect to their relevance to S can have the following formula

$$f(S, S_1, S_2) = p_1 p(S/S_1)^2 + p_2 p(S/S_2)^2 \tag{6.12}$$

where p_1 and p_2 are proportions of S_1 and S_2 in S_w, and $p(S/S_1)$ and $p(S/S_2)$ are proportions of the target category S in split parts S_1 and S_2, respectively. This formula follows from expressions (5.20) and (5.21) in section 5.2.4 at $|V_l| = 1$ and $b_v = 1$ with category v being an attribute defining the target cluster S. Indeed, the value of the contribution (5.21), in the current denotations, is $p_1(p(S/S_1)-p(S))^2+p_2(p(S/S_2)-p(S))^2 = p_1p(S/S_1)^2-p(S)^2+p_2p(S/S_2)^2-p(S)^2$ which coincides with $f(S, S_1, S_2)$ up to the constant $-2p(S)$.

6.3.3 Describing a cluster with production rules

Branches of a decision tree leading to terminal nodes labeled by class S_k of the partition which is described by the tree, can be considered production rules of the format "if A then S_k" where A is the conjunctive predicate corresponding to an individual branch. Such a production rule states that if A is true for $i \in I$, than i must belong in S. However, if A is not true for i, this does not necessarily imply that i does not belong to S.

The problem of producing such production rules for a single cluster S_k, without any relevance to other clusters, has attracted considerable attention from researchers. The problem fits well into the context of describing single clusters rather than partitions.

A production rule technique produces a set of statements "if A_p then S" $(p \in P)$ at which A_p is a conceptual, typically conjunctive, description of a subset in the feature space, that is, a predicate. A more precise formulation of a production rule would be "for all x in the feature space, $A_p(x) \rightarrow x \in S$." Rectangles surrounding black circles on Figure 6.8 correspond to such production rules. A production rule is characterized by two numbers: (a) the support, that is the number of entities in the set I satisfying A_p, (b) the precision, the proportion of entities from S among those entities of I that satisfy A_p. In some methods, descriptions A_p may overlap [66], [126], in others they don't [15], [1]. As noted above, branches of a decision tree labeled by S can be considered conjunctive production rules as well, however techniques generating production rules directly are more specific, as they are focused only on S, the group of interest. Thus, there can be no false negatives from a production rule, only false positives.

Production rules may have rather small supports. Moreover, different production rules may cover different parts of S and leave some parts not covered as shown on Figure 6.8.

Production rule techniques conventionally have been developed for prediction rather than for description. A specific task of prediction of warehouse transactions has led to a set of the so-called association rule techniques comprising a different section of data mining [23], [44], [114]. A transaction database's entities, transactions, are described by a set of goods purchased by a customer so that each of the transactions is a set of items purchased in one go. An association rule is a production rule "if A then B" where A and B are nonoverlapping sets of goods. Presence/absence of an individual commodity is a binary feature. Thus, an association rule can be considered a production rule for binary data so that association rule techniques can be applied to the problem of a single cluster description by considering belongingness to S as B. A number of techniques have been developed for producing association rules (see a review in [23], p.164-192).

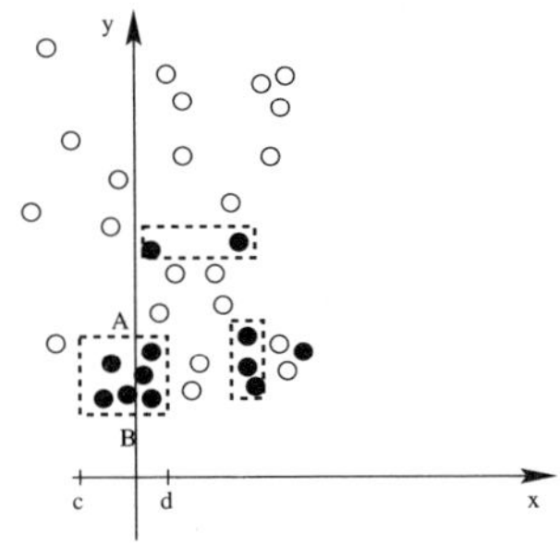

Figure 6.8: Rectangles in the feature space corresponding to production rules.

6.3.4 Comprehensive conjunctive description of a cluster

Methods for deriving what can be called a comprehensive description is a relatively recent addition to the machine learning literature. A comprehensive description of a subset $S \subset I$ is a statement "$x \in S$ if and only if $A(x)$" where $A(x)$ is a predicate defined for every entity in the feature space. Thus, a comprehensive description A is a unity of two production rules, "if $A(i)$ then $i \in S$" and "if $i \in S$ then $A(i)$."

Producing a comprehensive description is of interest in the situations in which S has been compiled in a process involving intuition and informal considerations, such as in the case of groups of protein folding or other microbiological taxonomy classes.

Correctness of a comprehensive description is characterized by the numbers of false positives and false negatives. A false positive for a description A of subset S is an element $i \in I - S$ satisfying $A(i)$, and a false negative is an element i from S, at which $A(i)$ is false. These errors correspond to the errors of the first and second kinds in the theory of statistical hypotheses if $A(x)$ is considered as a hypothesis about S.

A comprehensive description of S in the example of black and blank circles on Figure 6.9 is obtained by enclosing the subset S in a rectangle representing predicate $(a \leq x \leq b)\&(c \leq y \leq d)$.

Obviously, the problem of finding a unified comprehensive description of a group of entities can be solved by finding production rules predicting it and combining them disjunctively. However, when the number of disjunctive terms is constrained from above, in the extreme case just by 1, the problem becomes much more challenging especially if S is not "compact" geometrically but spread over the data set I in the feature space. Methods for finding a comprehensive description of a cluster, based on forward and backward search strategies have

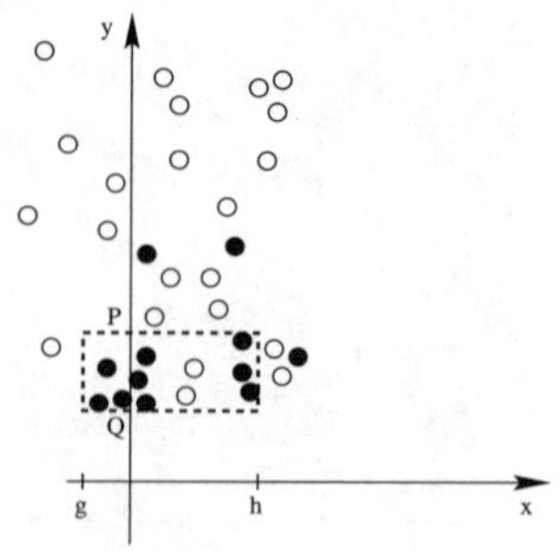

Figure 6.9: Rectangular description defining the numbers of false positives (blank circles within the box) and false negatives (black dots outside of the rectangle).

been considered in [92] and [107]. In [107] the difficulty of the problem was partly avoided by applying the techniques to cohesive clusters only and in [92] transformation of the space by arithmetically combining features was used as a device to reduce the spread of S over I.

An algorithm outlined in [92] has the following input: set I of entities described by continuously-valued features $v \in V$ and a group $S \subset I$ to be described. The algorithm operates with interval feature predicates $P_f(a,b)$ defined for each feature f and real interval (a,b), of values $a \leq x \leq b$: $P_f(a,b)(x) = 1$ if the value of feature f at x falls between a and b, and $P_f(a,b)(x) = 0$ otherwise. The output is a conjunctive description of S with the interval feature predicates along with its false positive and false negative errors.

The algorithm involves two building steps:

1. **Finding a conjunctive description in a given feature space V.** To do this, all features are initially normalized by their ranges or other coefficients. Then all features are ordered according to their contribution weights which are proportional to the squared differences between their within-group $S \subset I$ averages and grand means, as described in section 3.4.2. A conjunctive description of S is then found by consecutively adding feature interval predicates $f_v(a,b)$ according to the sorted order, with (a,b) being the range of feature v within group S ($v \in V$). An interval predicate $f_v(a,b)$ is added to a current description A only if it decreases the error.[1] This forward feature selection process stops after the last element of V is checked. Then a backward feature search strategy is applied to decrease the number of conjunctive items in the description, if needed.

[1] In this way, correlated features are eliminated without much fuss about it.

2. **Expanding the feature space V.** This operation is applied if there are too many errors in the description A found on the first step. It produces a new feature space by arithmetically combining original features $v \in V$ with those occurring in the found description A.

These two steps can be reiterated up to a pre-specified number of feature combining operations. Step 2 transforms and arithmetically combines the features to extend the set of potentially useful variables. Then Step 1 is applied to the feature space thus enhanced. The combined variables appearing in the derived decision rule are then used in the next iteration, for combining with the original variables again and again, until a pre-specified number of combining operations, t, is reached. For example, one iteration may produce a feature $v1 * v2$, the second may add to this another feature, leading to $v1 * v2 + v3$, and the third iteration may further divide this by $v4$, thus producing the combined feature $f = (v1 * v2 + v3)/v4$ with three combining operations.

This iterative combination process will be referred to as APPCOD (APProximate COmprehensive Description). An APPCOD iteration can be considered a specification of the recombination step in genetic algorithms [99]. The difference is that the diversity of the original space here is maintained by using the original variables at each recombination step, rather than by the presence of 'bad' solutions in conventional genetic algorithms.

In this way, given a feature set V, entity set I, and class $S \subset I$, APPCOD produces a new feature set $F(V)$, a conjunction of the interval feature predicates based on $F(V)$, A, and its errors, the numbers of false positives FP and false negatives FN. Since the algorithm uses within-S ranges of variables, it fits into the situation in which all features are quantitative. However, the algorithm can also be applied to categorical features presented with dummy zero-one variables; just intervals (a, b) here should satisfy the condition $a = b$, that is, correspond to either of values, 1 or 0.

It should be mentioned that a more conventional approach to finding a good description A with mixed scale variables involves building logical predicates over combinations of features [24] and [69].

Example 6.51. Combined features to describe Body mass groups

Let us apply APPCOD to the Body mass data in Table 1.7, with the restriction of no more than one feature combining operation. Depending on which of the groups, overweight or normal-weight, is to be described, the algorithm first selects either Weight or Height. This gives a lousy description of the group with too many false positives. Then APPCOD combines the feature with the other one and produces the difference Height − Weight and the ratio Height/Weight for both the overweight and the normal-weight groups. The overweight group is described with these combined features with no errors, whereas the normal group cannot be described distinctively with them. This means that the former group is somewhat more compact in the combined feature space. The overweight group's comprehensive description is: $2.12 \leq H/W \leq 2.39$ & $89 \leq H - W \leq 98$, with no errors.

The second inequality can be reinterpreted as stating that the difference Height – Weight for the other, normal group, is about 100, which fits well into the known common sense rule: "Normally, the difference between Height (cm) and Weight (kg) ought to be about 100."

When the number of permitted feature combining operations increases to 2 and the number of conjunctive items is limited by 1, the method produces the only variable, $H * H/W$, which distinctively describes either group and, in fact, is the inverse body mass index used to define the groups. □

6.4 Overall assessment

There is a great number of clustering methods that have been and are being developed. In section 6.1, a number of different clustering methods that can be treated as extensions of K-Means clustering have been presented. These methods are selected because of their great popularity. There are rather straightforward extensions of K-Means among them, such as Partitioning around medoids, in which the concept of centroid is specified to be necessarily an observed entity, and Fuzzy K-Means, in which entity memberships can be spread over several clusters. Less straightforward extensions are represented by the regression-wise clustering, in which centroids are regression functions, and self-organizing maps SOM that combine the Minimum distance rule with visualization of the cluster structure over a grid. The probabilistic mixture of distributions is a far reaching extension, falling within the statistics paradigm with a set of rather rigid requirements such as that all features are continuous-valued and must be standardized with the z-score transformation, none of which needs to be assumed within the data recovery approach.

Two graph-theoretic clustering methods are presented in section 6.2. One is the quite popular single linkage method related to connected components and minimum spanning trees. The other is a rather new method oriented towards finding a central dense core cluster; the method works well when such a dense core is unique.

Conceptual description of clusters is covered in the last section. Three different types of conceptual description are highlighted: (1) decision tree, the most popular means for conceptual classification in machine learning, (2) production rule, a device for picking up conditions leading to belongingness to a cluster, and (3) comprehensive description, which is oriented towards formulation of necessary and sufficient conditions of belongingness to a cluster. The problem of conceptual description of clusters has not yet found any general satisfactory solution.

Chapter 7

General Issues

After reading through this chapter you will get a general understanding of existing approaches to:

1. Data pre-processing and standardization;
2. Imputation of missing data;
3. Feature selection and extraction;
4. Various approaches to determining number of clusters;
5. Validation of clusters with indexes and resampling;

and contributions of the data recovery approach to these issues.

Base words

Bootstrapping A resampling technique in which a copy of the data set is created by randomly selecting either rows or columns, with replacement, as many times as there are respective number of rows or columns in the data.

Cluster validation A procedure for validating a cluster structure or clustering algorithm. This can be based on an internal index, or external index or resampling. An internal index scores the degree of correspondence between the data and the cluster structure. An external index compares the cluster structure with a structure given externally. A resampling is used to see whether the cluster structure is stable with respect to data change.

Cross-validation A resampling technique that provides for a full coverage of the data set. The data set is randomly divided in a number Q of equally-sized parts and Q copies of the data in the 'training/testing' format are generated by taking each of the parts as the test part in a corresponding copy.

Dis/similarity between partitions A partition-to-partition measure based on the contingency table. All measures reviewed in this chapter can be expressed as the weighted averages of subset-to-subset measures.

Dis/similarity between sets A subset-to-subset measure relying on the subsets' overlap size. A popular Jaccard coefficient relates the overlap size to that of the union and has both advantages and shortcomings. Somewhat better measures should involve the ratios of the overlap size to each of the subsets or Quetelet coefficients.

Feature extraction A procedure for creating such 'synthetic' data features that maintains the most important information of a pattern in question.

Feature selection A procedure for finding such a subspace of data features that retains the most important information of a pattern in question.

Index based validation Validation of a cluster structure involving either an internal index, measuring degree of correspondence between the data and the clusters, or an external index, measuring the correspondence between the cluster structure and an externally specified partition or other external data.

Number of clusters An important characteristic of a cluster structure, which depends both on the data and the user's objectives. There have been developed a number of criteria for determining a 'right' number of clusters, both index-based and resampling based; so far none has come up as applicable across diverse data ranges.

Resampling An approach to creating an empirical distribution of a data mining pattern or criterion by creating a number of random copies of the data table and independently applying to them the method for computing the pattern or criterion of interest. The distribution is used for assessing validity of the algorithm or pattern/criterion.

7.1 Feature selection and extraction

7.1.1 A review

The problem of reducing the number of features may arise in a technical context to make a huge data set treatable computationally. In some contexts the driving force can be interpretation, in which the number of features involved seems a good index of clarity: the better a problem or phenomenon is understood the fewer the number of features needed to describe it. Issues of reducing the size of the feature space in clustering problems attract considerable interest from researchers [18, 76].

There are several dividing lines apparent in these efforts. One is to distinguish between methods for feature selection and methods for feature extraction. The former are aimed at choosing a smaller subset from the entire set of features and the latter at producing a few 'synthetic' variables from the given ones. Another divide is between situations in which clusters are pre-specified and those in which they are not, that is, between supervised and unsupervised learning problems.

One more division arises from the way selected features are assessed. In particular, under supervised learning, when a partition is pre-specified, two types of methods of feature selection or extraction are distinguished [18]: *wrapper* and *filter*. When a method for class prediction, a classifier, has been selected, its results on different feature spaces can be compared with the pre-specified partition and, in this way, used for selection of the best feature space. Comparison of the classifier results with the partition can be done according to different scoring functions. The simplest is just counting the number of errors made by the classifier on a selected feature set; the fewer errors the better. More complex measures use various weightings of errors/correct predictions depending on the partition class sizes; a review of some comparison measures can be found in [137]. Since the interrelation between the prespecified partition and wrapper results amount to a confusion matrix, which is a contingency table, contingency association coefficients discussed in sections 2.2.3 and 3.4.4 can be utilized, as well as the external validation indexes in section 7.5.1. The classifier serves as a wrapper of different feature spaces, which explains the name given to this type of evaluation methods. Wrapper methods can be used for both feature selection and feature extraction. Results of a wrapper method are straightforward and easy to use; however, they may depend highly on the wrapper utilized.

To apply wrapper methods for feature selection in clustering, even for a situation in which a 'true' partition has been pre-specified, a clustering, rather than prediction, method should be used as a wrapper. Such a wrapper clustering method can be referred to as a cluster maker. The closer clustering results are to the true partition, the better the cluster maker. Obviously, the same strategy can be applied when the quality of a feature space is to be judged by the relation of results not necessarily to a pre-specified partition but some other feature(s).

Another class of methods constitute so-called filter methods. A filter method selects or builds features based on a scoring function $f(W)$ defined for every feature set $W \subseteq V$.

Probably the simplest within this class is the case of additive $f(W)$ in which $f(W)$ can be represented as the sum of feature weights, $f(W) = \sum_{v \in W} f_v$, where f_v is feature v's weight. Under supervised learning, when a pre-specified partition is known, feature weights can be defined according to their association with the partition or its classes. In particular, for a quantitative feature, its cluster-specific weight is typically defined by the difference between the feature's grand mean and within cluster mean: in general, the greater the difference the further away is the cluster from the data center, thus the better the feature for separating the cluster from the rest [66]. For a categorical feature, its salience depends on its association to the partition. This means that contribution coefficients derived in section 5.2 are appropriate for the task. Sometimes, feature weights are defined within a classifier computed with respect to the pre-specified partition, as, for example, standardized coefficients of variables in a linear discriminant function or logistic regression.

In the unsupervised situation feature saliencies can be defined according to feature variance measured by contribution to the scatter, or entropy, or dissimilarity from other features. For example, according to a traditional approach, the feature set is first partitioned into groups of similar features, and then a 'central' feature is selected from each of the groups. Such a method will produce feature subset $\{NC, SC, Pe\}$ based on the partition of the Masterpieces' feature set presented in Figure 6.6 (b).

Sometimes a scoring function can be defined according to substantive considerations. For instance, in text mining features are key words or phrases; the number of occurrences of a keyword appears to be a good indicator: the most relevant keywords are those whose occurrence numbers are somewhere between the minimum (occasional terms) and maximum (common words). More on weighting of variables can be found in [37].

If the scoring function $f(W)$ is not additive, which is the case in wrapper methods, its optimization can be computationally challenging. To simplify calculations, greedy methods are frequently employed. Among them the most popular are so-called Forward Search Selection (FSS) and Backward Search Selection (BSS) approaches. An FSS method starts with empty W and iteratively updates W by adding just one, the best possible, feature at each iteration until a stopping criterion is met. A stopping criterion may involve reaching a pre-specified number of features in W or a threshold to the value of $f(W)$. A BSS method starts from $W = V$, the set of all features, and at each iteration removes one feature v from it; that one at which $f(W - v)$ is optimal over $v \in W$.

7.1.2 Comprehensive description as a feature selector

Let us concentrate on the task of learning a subset $S \subset I$. A filter method can be employed for selecting the most salient features. The salience of feature v with respect to S should be measured based on the ability of v to separate S from the rest. To this end, the difference between v's grand mean and within-S mean can be utilized. In particular, its squared value c_{Sv}^2 can be used as pertaining to the contribution of feature v at S to the data scatter (see section 3.4.2). With the data preliminarily pre-processed by shifting them to the grand mean a_v and normalizing by b_v, $c_{Sv}^2 = (a_v - a_{Sv})^2/b_v^2$ where a_{Sv} is the within-S mean of feature v. Note that both a_v and a_{Sv} are calculated at the original variable x_v. When x_v is a zero-one binary feature corresponding to a category, $c_{Sv}^2 = (p_v - p(v/S))^2/b_v^2$ where p_v and $p(v/S)$ are unconditional and conditional frequencies of v. Denoting the proportion of S in I by p_S and proportion of the overlap of v and S by p_{Sv}, this can be further reformulated as $c_{Sv}^2 = (p_{Sv} - p_v p_S)^2/p_S^2$ or $c_{Sv}^2 = (p_{Sv} - p_v p_S)^2/(p_v p_S)$ at $b_v = 1$ or $b_v = \sqrt{p_v}$, respectively.

Selecting features with largest c_{Sv}^2 is computationally effective and simple. However, features picked this way can be highly correlated, thus reducing their usefulness. Using methods of cluster description described in section 6.3 to post-process salient features can mitigate the issue. In particular, the APPCOD method from section 6.3.4 can be used as a feature selecting device: the set of selected features W is formed by those features that are involved in the comprehensive description of S found with APPCOD.

Example 7.52. APPCOD based feature selection applied to Iris classes

In the Iris data set, predefined classes can be described by the following concepts found with the algorithm APPCOD: $1 \leq w3 \leq 1.9$ (class 1, FP=0), $3.0 \leq w3 \leq 5.1$ & $1.0 \leq w4 \leq 1.8$ (class 2, FP=8), and $1.4 \leq w4 \leq 2.5$ & $4.5 \leq w3 \leq 6.9$ (class 3, FP=18). Since the method APPCOD uses within-class ranges of features for producing interval predicates, numbers of false negatives FN, in this version, are all zero. Numbers of false positives FP at conjunctions describing classes II and III are rather high, but they cannot be reduced by adding other variables' ranges. High errors for two of the Iris classes support the conclusion that they are dispersed in the variable space (see Figure 1.10, page 25).

Still, only two variables occur in the descriptions obtained for the Iris classes, $w3$ and $w4$. This effectively selects these two variables as those that matter. □

Obviously, a conceptual description of a cluster even when it is precise is not necessarily meaningful from the substantive point of view: measured features may be not quite essential to the process underlying differences between clusters. There is no cure-all remedy to this effect within data mining. If, for instance, some patients suffer from age-related ailments, it is highly unlikely that a transformation of location and income related variables can possibly be a good substitute for a direct age characteristic.

7.1.3 Comprehensive description as a feature extractor

In the early days of the development of multivariate data analysis methods, feature extraction was confined to extracting 'hidden' factors being linear combinations of the original variables. This is still a niche for Principal component analysis (PCA) and related methods described in sections 5.1.3 and 5.5.1, which have been recently grossly enhanced by the advent of kernel based methods [36, 110]. In this context, kernel based methods provide for non-linear transformations of the original features to effectively linearize the task of supervised learning. An issue with all these methods is of an artificial character of the 'synthetic' features emerging, which is rather disappointing in the context of clustering because new features are needed mainly for better understanding.

More recently, a number of methods emerged, especially in supervised learning, that combine original features to produce meaningful descriptions of classes [92, 107]. APPCOD method[92] described in section 6.3.4 seems rather convenient in this regard as it uses arithmetic combinations of original features, which nicely fits into the long-standing tradition of sciences. According to this tradition, derivative measures such as area or electric current can be expressed as arithmetic products or ratios of other features such as length and width or voltage and resistance.

Example 7.53. APPCOD based feature extraction applied to Iris classes
Let us try APPCOD as a feature-extracting device on the Iris data set. Assume that features w3 and w4, selected with APPCOD in the previous section, constitute set A which will be arithmetically combined with the four original Iris variables in set V. We denote by $F(A, V)$ the set of features produced by applying each of five operations $(x+y, x-y, x*y, x/y, y/x)$ to each $x \in A$ and $y \in V$. APPCOD applied to $F(A, V)$ for description of Iris classes II and III (class I has been distinctively separated by $w3$ alone and thus excluded from further analysis), produces conjunctions $1.18 \leq w1/w3 \leq 1.70$ & $3.30 \leq w3 * w4 \leq 8.64$ (class II, FP=4) and $7.50 \leq w3 * w4 \leq 15.87$ & $1.80 \leq w3 - w2 \leq 4.30$ (class III, FP=2) with the number of conjunctive terms restricted to being not greater than 2. The errors have become smaller, but still they may be considered too high. Can the numbers of false positives be reduced to just one per class?

Putting the compound variables involved in the descriptions above, $w1/w3$, $w3 * w4$, and $w3 - w2$, as A and leaving V as is, an update $F(A, V)$ can be computed to give rise to the following APPCOD produced conjunctions: $2.86 \leq w1 * w2/w3 \leq 4.77$ & $3.30 \leq w3 * w4^2 \leq 15.55$ (class II, FP=4) and $3.24 \leq (w3 - w2) * w4 \leq 9.89$ & $1.35 \leq w3 * w4 - w1 \leq 8.70$ (class III, FP=2).

The errors of the two-term conjunctions have not changed. However, one could see that the new feature space leads to a better four-term conjunctive description of class III (with FP decreased to 1). With set A consisting of the four new variables, $w1 * w2/w3$, $w3 * w4^2$, $(w3 - w2) * w4$, and $w3 * w4 - w1$, and V unchanged, the algorithm APPCOD applied to $F(A, V)$ leads to the following conjunctions: $0.64 \leq w2 * (w3 - w2) * w4 \leq 4.55$ & $0.21 \leq w2/(w3 * w4^2) \leq 0.74$ (class II, FP=1) and $4.88 \leq w3 * w4^2 - w1 \leq 31.20$ & $-2.85 \leq (w3 - w2) * w4 - w1 \leq 2.19$ (class III, FP=1). It should be added that class I can be distinctively separated with one of these variables, $w2 * (w3 - w2) * w4 \leq -3.07$ (class 1, FP=0).

Table 7.1: Confusion matrix for iK-Means clusters, in the original feature space, and predefined classes of the Iris data.

Classes	iK-Means clusters, option 1						iK-Means clusters, option 2					
in Iris data	1	2	3	4	5	6	1	2	3	4	5	Total
1	0	49	1	0	0	0	0	50	0	0	0	50
2	0	0	13	10	5	22	0	0	15	16	19	50
3	12	0	2	18	18	0	27	0	3	21	0	50
Total	12	49	16	28	23	22	27	50	18	36	19	150

Having achieved the goal of not more than one error per class, the process of combining variables is stopped at this point to underscore a trade-off between the exactness and complexity of cluster descriptions, which parallels similar trade-offs in other description techniques such as regression analysis. The final features do not make much sense. However, an intermediate feature emerged and conserved, $w3 * w4$, referring to the petal area, is considered by many as a key feature to substantively discriminate between Iris genera II and III. □

Example 7.54.

Extracting Body mass features with APPCOD

Example 6.51 of using APPCOD for separating two classes of Body mass data can be considered in the feature extracting context, too. Depending on what group is considered as S, those of overweight or normal weight, the method leads to either Height − Weight or Height/Weight^2, as an extracted combined feature. Both of these features are known to be quite meaningful in the problem of separation of the overweight from the normal. □

As explained above, criteria for evaluation of the quality of extracted features can be based on the performance of a classifier or cluster maker in the new feature space. In the clustering context the latter seems preferable.

Example 7.55.

Quality of feature spaces according to iK-Means

Let us explore the quality of different feature spaces produced in the previous examples to tackle the issue of supervised feature selection and extraction on the Iris data set. Let us utilize the intelligent version of K-Means, iK-Means described in section 3.3, as a cluster maker.

First, let us consider results of iK-Means applied to the original data set with two standardizing options: (1) features z-score standardized; (2) features range standardized, that are presented in the left and right contingency tables in Table 7.1, respectively. The cluster discarding threshold in iK-Means was set to be equal to 2.

Table 7.1 shows that the level of confusion is less in the case of the range standardized data (option 2). This is caused by the fact that with this standardization, contributions of features to the data scatter are shifted towards those petal related, w3 and w4, which better correlate with the genera, as was shown in the example 7.52: the vector of contributions of features w1-w4 under option (2) is $(19, 12, 32, 37)$, per cent, whereas all contributions are equal to 25% for the z-score standardization.

The levels of confusion are lessened even more if we impose the restriction that

Table 7.2: Confusion matrix at iK-Means clusters set to 3 (in all three options) and predefined classes of the Iris data.

Classes	Clusters, op. 1			Clusters, op. 2			Clusters, op. 3			
in Iris data	1	2	3	1	2	3	1	2	3	Total
1	0	49	1	0	50	0	0	50	0	50
2	13	0	37	10	0	40	2	0	48	50
3	42	0	8	42	0	8	46	0	4	50
Total	55	49	46	52	50	48	48	50	52	150

Table 7.3: Confusion matrix of iK-Means clusters, in the extracted feature space, and predefined classes of the Iris data.

Classes	Clusters, 1				Clusters, 2			
in Iris data	1	2	3	4	1	2	3	Total
1	50	0	0	0	50	0	0	50
2	0	2	41	7	0	2	48	50
3	0	48	2	0	0	48	2	50
Total	50	50	43	7	50	50	50	150

the number of clusters in iK-Means must be the same as the number of genera, that is, 3, as clearly seen in Table 7.2. One more option, 3, is added here, with the feature set confined to only two features, w3 and w4, that have been selected by APPCOD in example 7.52. This last option leads to a solution with only 6 displaced entities, which supports the principle of feature selection.

Let us take a look now at clustering results with extracted features. Table 7.3 presents results of iK-Means applied to the APPCOD enhanced features w1/w3, w3*w4 and w3-w2 from the previous example. The method leads to a four cluster solution presented in the left part (option 1); the right part presents the solution found by iK-Means restricted to having 3 clusters only (option 2). The range standardized features contribute almost equally to the data scatter in this case, which implies that results would not differ under the z-score based data standardization.

All the confusion in the extracted feature space is created by four specimens, 13 and 32 of class II, and 33 and 39 of class III, since they are still closer to the other class centroid than to that of the class they belong to. It is worth adding that the separation of cluster 2 is mostly due to the contribution of the "area" feature w3*w4, while cluster 3 is overwhelmingly supported by the ratio of sepal length to petal length w1/w3.

Results in Table 7.3 are the best we could obtain with APPCOD-extracted features of the previous example. Contrary to expectations, further combining features does not improve the quality of clustering, probably because more complex features relate to finer details of differences between entities, which do not help in separating the genera but rather break them up according to the finer granulation. Let us take a look, for example, at iK-Means clustering found at the final feature set ob-

Table 7.4: Confusion matrix for iK-Means clusters, in a complex feature space, and predefined classes of the Iris data.

Classes	iK-Means clusters							
in Iris data	1	2	3	4	5	6	7	Total
1	0	5	45	0	0	0	0	50
2	0	0	0	0	22	7	21	50
3	15	0	0	25	0	10	0	50
Total	15	5	45	25	22	17	21	150

tained using three combining operations in the previous example, w2*(w3-w2)*w4, w2/(w3*w4*w4), w3*w4*w4-w1, and (w3-w2)*w4-w1 (Table 7.4).

Five entities of the first genus are extracted as cluster 2 here because they have by far the greatest values of feature w2/(w3*w4*w4), which is also responsible for the emergence of a partial cluster 7. This delicate feature, along with (w3-w2)*w4-w1, leads to the production of the mixed cluster 6 as well. □

7.2 Data pre-processing and standardization

7.2.1 Dis/similarity between entities

Clustering algorithms are defined over sets of entities at which a dis/similarity measure has been or can be defined. We use here term dis/similarity to express both types of between entity proximity measures, dissimilarities and similarities. Dissimilarities, typically, are non-negative reals: the closer are entities to each other the smaller dissimilarities are, decreasing to 0 to express the identity case. In contrast, similarities can be negative and they are increased to express closer ties between entities.

A (weighted) graph can be considered a similarity matrix on the set of its vertices, with similarity s_{ij} equal to the weight assigned to the edge joining vertex i with vertex j; in the ordinary graphs, s_{ij} is 1 or 0 depending on whether the arc from i to j exists.

In the collection presented in section 1.1, Confusion represents similarity data and Primates dissimilarity data. Euclidean squared distance and inner product are important examples of dissimilarity and similarity measures derived from feature-based data.

An inversely monotone transformation such as $f(s) = A - s$ or $f(s) = e^{-\alpha s^2}$ or $f(s) = \frac{1}{1+s^2}$ transforms a similarity measure s into a dissimilarity measure $f(s)$ and vice versa.

Some may claim that the dis/similarity format is indeed the only one to formulate clustering concepts and algorithms because any cluster, in the end, is determined in terms of dis/similarities between entities. In fact, features can be useful as well because:

1. Clustering methods such as K-Means may employ the specificity of the vector space format to explicitly formulate the concept of cluster's prototype or centroid as a feature based entity.

2. Features are involved in cluster interpretation.

3. The vector space format can be by far more effective computationally when the number of features is relatively small. Think of the difference between NM input numbers in the vector space and N^2 or $N^2/2$ input numbers in the dis/similarity format when N is large and M small. If, for instance, there are 10,000 entities and 100 features, there will be about one million numbers in the former case and about a hundred million numbers in the latter case, the difference which can be crucial with contemporary personal computers.

The idea of conversion of dis/similarity based data into the feature space format motivated a lot of research in the so-called multidimensional scaling, the discipline oriented towards embedding given dis/similarity data into a vector space in such a way that the between-entity distances in the space would approximate the prespecified dis/similarities. An issue with this approach is that the found space has no intrinsic interpretational support. In our view, this can be overcome with another idea gaining more and more popularity that, given a dis/similarity data table, a number of entities should be pre-selected as reference points serving as coordinate axes so that each entity can be represented by the vector of its distances to the reference points.

Complex data such as chemical compound formulas, time-series of stock prices, biomolecular sequences, images, etc. can be used for automatically discovering patterns such as motives in sequences or edges in images that can be used then for feature generation. This implies that the pre-processing issues are not necessarily purely computational ones but can involve specific data mining techniques including clustering. These topics remain beyond the scope of this text. Relevant materials can be found in [23] (sequences and temporal data), [123] (images), [44] (data bases and warehouses), [26] (spatial data).

7.2.2 Pre-processing feature based data

The only pre-processing issue raised in this book concerns categorical and mixed scale features.

Traditionally, mixed scale feature data are treated by transforming them into a dissimilarity matrix $D = (d_{ij})$, where $i, j \in I$ are entities, according to thc formula:

$$d_{ij}^p = \sum_{f \in F} w_f |x_{if} - y_{jf}|^p + \sum_{v \in V} w_v |x_{iv} - y_{jv}|^p \tag{7.1}$$

where p is the so-called Minkowsky's power, typically, taken as $p = 2$ or $p = 1$; F, V are respectively sets of quantitative features and qualitative categories; w_f and w_v are user-defined weight coefficients for features $f \in F$ and categories $v \in V$; and x_{iv} are 0/1 values (1 at entities i falling in the category v and 0 at the rest). Similarity measures are defined analogously by summing up weighted similarities between i and j at individual features and categories.

Some may say that formula (7.1) does not much differ, at $p = 2$, from the Euclidean distance squared used in this text, and in fact, is even more general because it involves weights w_f and w_v. Yes, indeed. It is exactly the extent of generality which is fought off with the help of the data recovery approach. The common opinion is that the weight coefficients are to come from the user. "The choice of measure will be guided largely by the type of variables being used and the intuition of the investigator" ([28], p. 54). The current author argues this because in most cases the user has no clue regarding to the weights in formula (7.1). According to this book's approach the weights are supposed to come from data, not the user. This is provided by the recommended pre-processing and standardization technique described in section 2.4. With data x_{iv} and x_{if} standardized into y_{iv} and y_{if} the weights in distance (7.1) become $w_v = 1/b_v^p$, $v \in F \cup V$, with b_v being scaling coefficients derived from the data.

To substantiate this recipe, let us bring forward the following:

1. Introducing dummy variables for categories extends the concept of quantitative scale as that admitting the operation of averaging to qualitative categories: averaging dummies leads to conditional and unconditional probabilities (frequencies), which is exactly the machinery utilized in methods specifically oriented for analysis of categorical data.

2. Rescaling of categories is an effective instrument to maintain appropriate contributions of categories to the data scatter and, in this way, provide consistency to the process of simultaneously analyzing quantitative and qualitative data.

3. Summary contributions of dummy variables to the explained part of the data scatter appear to be in line with intuitive criteria proposed within the qualitative data framework such as the category utility function and Pearson chi-square contingency coefficient applied in conceptual clustering and decision tree building. This property brings a rather different light to the contingency coefficients for cross classifications. It appears,

they must be considered on par with the correlation ratio for quantitative variables and, moreover, get unexpectedly transformed, one into the other, depending on the data normalization option accepted.

4. The data recovery framework helps in addressing one of the major issues of data analysis: warranting that data pre-processed into the dis/similarity format get no additional structures imposed by the pre-processing. Statements proven in sections 5.4 and 5.5 allow us to claim that there are criteria and methods that should lead to compatible results in both feature-based and dis/similarity data. More explicitly, if a dissimilarity measure is equal (or similar) to the squared Euclidean distance squared between rows of a vector space matrix, or a similarity measure is the row-to-row inner product, then compatible criteria and local search heuristics are those detailed in section 5.4.1.

7.2.3 Data standardization

Data standardization is performed feature-wise and/or entity-wise to make features and/or entities comparable. The issue of comparability of features can be illustrated with the Market towns data set in Table 1.1: obviously the scale of feature Population is a thousand times greater than of any other feature in the table, thus differences in this feature will dominate the dis/similarities between entities. This contradicts the *principle of equal importance of features*, which underlies cluster analysis as well as many other methods of multivariate data analysis. The principle states that in the situation when no explicit weights can be assigned to features they should equally contribute to the result. This principle is underlined by the assumption that the user controls the process of data analysis by structuring the phenomenon under consideration into meaningful aspects and then assigning features to the aspects. The relative importance of an aspect is reflected by the number of features assigned to it. In this way, the equivalence of features is translated into the relative importance of the aspects.

The practical implementation of the equivalence principle is usually done via, first, standardization of the variables and, second, balancing contributions of standardized features into entity-to-entity distances [122, 58]. A similar approach has been advised in this text. The only difference is that the distances, in our context, are not a primary but derivative tool: it is the data recovery models and criteria which are primary. The distances are generated by the square error criteria of these models; the balance of contributions can be maintained in the criteria as well, and data scatter decompositions provide a feedback to compare contributions.

Shifting the origin and changing the scale with formula $y_{iv} = (x_{iv} - a_v)/b_v$ (2.22) is the main device for feature standardization. More radical proposals include further data transformation along lines that are akin to the singular value decomposition of the data matrix and, thus, should be considered part of data analysis rather than data pre-processing. It should be pointed out that there are cases in which the scale may be left unchanged, that is, b_v taken to be unity: such is the case of binary features or, more generally, the situations in which every feature presents a ranking within the same rank scale, say from 1 to 5.

However, proposals for setting values of a_v and b_v may highly differ. In particular, the value of the shift coefficient a_v has been suggested in the literature to be equal to any of the following:

(a1) 0;

(a2) minimum, $\min_{i\in I} x_{iv}$;

(a3) maximum, $\max_{i\in I} x_{iv}$;

(a4) midrange, $(\max_{i\in I} x_{iv} + \min_{i\in I} x_{iv})/2$;

(a5) grand mean, the average of x_{iv} over all $i \in I$;

(a6) median.

Similarly, it has been suggested that the value of the scale coefficient b_v be taken as any of the following:

(b1) range r_v or half-range $r_v/2$ with $r_v = \max_{i\in I} x_{iv} - \min_{i\in I} x_{iv}$;

(b2) standard deviation $s_v = \sqrt{\sum_{i\in I}(x_{iv} - \bar{x_v})^2/N^*}$ where N^* is either N or $N-1$, the latter value due to probabilistic considerations of mathematical statistics;

(b3) grand mean $\bar{x}_v$, the average of x_{iv} over all $i \in I$.

Among scaling coefficients, the most popular option is the standard deviation, probably because it is inherited from methods of mathematical statistics assuming the normal distribution of data at which the joint action of options (a5) and (b2), referred to as z-scoring, is the natural choice. I also subscribed to this option, in the context of non-probabilistic data analysis, as an explication of the principle of equal importance of features: with $b_v = s_v$, contributions of all features to the data scatter are equal to each other [90]. However, this recommendation is overly simplistic. The value of standard deviation depends on two factors: the feature scale size and the shape of its distribution. Obviously, the distribution shape is important in clustering: a multi-modal distribution may be evidence of the multi-cluster structure. However, normalizing features with the standard deviation forces bi- and more-modal features to shrink to a smaller weight than they should (see also section 2.1.2). The other two options, normalizing by the range or average, catch the size factor only, while leaving the distribution aside and, thus, should be preferred in clustering. A popular option rescales data in such a way that all values fall within interval [−1, 1], which is achieved with a_v being mid-range (a4) and b_v range (b1). Another option pulls all data into the interval [0,1], by using (a2) and (b1), to be able, formally, to apply operations and interpretations referring to proportions and frequencies.

In [85] a set of computational experiments have been performed with a number of distance-based clustering methods and various standardization options. The methods included a range of agglomerative procedures from the single linkage to Ward agorithms. The data generated consisted of a small number, 50, of points of small, 4 to 8, dimensions, concentrated in a number of well separated clusters, supplemented with errors being either (i) normally distributed noise added to data entries, or (ii) two additional, randomly generated, features, or (iii) 20% outlier points added to the set. Overall, the normalization by range appeared to unanimously provide for the best recovery of the underlying cluster structure among other normalization options; "the traditional z-score formula was not especially effective." ([85], p. 202.) This concurs with the advice following from the analysis of feature contributions to the Pythagorean decomposition in section 3.4.

Subtracted shift values a_v play no role in the distance (7.1), but they do work when using the inner product as a similarity measure and, more importantly, in cluster centroids. In this regard, the recommended option (a5) for a_v being the grand mean has obvious advantages. With this, the cluster centroids express the differences between the feature within-cluster averages and grand means – these differences accentuate each cluster's specifics. The differences are indispensable in interpreting clusters and, moreover, they are behind the entire system of cluster-feature contributions to the data scatter. The appropriateness of using grand means as the shift coefficients is underscored by the properties of these contributions relating them to classical ideas of statistics: the correlation ratio, Quetelet indexes, and the Pearson chi-square association coefficient. The working of the Anomalous pattern method and iK-Means clustering is based on that, too.

7.3 Similarity on subsets and partitions

Subsets and partitions play an important role in clustering on both sides of the data pipeline, the input and output.

On the input side, a binary category can be equivalently represented, up to the corresponding meta-data, by the subset of entities falling under the category. Thus, binary data tables can be considered sets of subsets. Binary features constitute a most important specific data format to which many other data formats can be pre-processed. Examples include:

1. Sets of keywords, expert-driven or found automatically, for documents;
2. Sets of neighboring representatives;
3. Genome contents;
4. Network neighborhoods.

In some applications, binary data are pre-processed to take into account their row and column frequencies. In particular, in information retrieval, the so-called tf-idf ("term frequency, inverse-document frequency") weighting scheme is widely accepted. This scheme assigns every unity in a document-to-keyword data table the product of the keyword frequency within the document and the logarithm of the proportion of total documents to the number of documents in which the term appears [116]. Another pre-processing option would be to treat a binary data table as a contingency table and code every entry by the corresponding Quetelet index defined in section 2.2.3. These transformations make the data not binary anymore and are not considered further on.

Partitions correspond to nominal features and, thus, also constitute an important class of input data.

On the output side, subsets and/or partitions are results of many clustering algorithms. To compare results of different classification schemes or clustering algorithms, one needs to measure similarity between them.

There have been a number of different approaches and measures of dis/similarity between subsets or partitions developed in the literature of which some will be reviewed here (see [56] and [90] for more).

7.3.1 Dis/similarity between binary entities or subsets

Set theoretic similarity measures

When all features are binary, every entity $i \in I$ can be considered as just a set (bag) of features V_i that are present at i. Then traditional set-theoretic operations can be used to express such concepts as "set of features that are absent from i" ($\bar{V_i}$), "set of features that are present at i but not at j" ($V_i - V_j$), "set of features that are common to i and j" ($V_i \cap V_j$), etc.

The so-called four-fold table presented in Table 7.5 is a traditional instrument for comparing subsets V_i and V_j in I. It is a contingency table cross classifying V_i and its complement $\bar{V_i} = I - V_i$ with V_j and its complement $\bar{V_j} = I - V_j$. Its interior entries are cardinalities of intersections of corresponding sets and its marginal entries are cardinalities of the sets themselves.

Table 7.5: Four-fold table.

Set	V_j	$\bar{V_j}$	Total
V_i	a	b	a+b
$\bar{V_i}$	c	d	c+d
Total	a+c	b+d	$a+b+c+d=N$

Probably the simplest similarity measure is the size of the overlap,

$$ij = |V_i \cap V_j| = a. \tag{7.2}$$

However, this measure fails to relate the size of the overlap to the set sizes and thus does not allow us to judge whether the overlap comprises the bulk or just tip of the sets. The distance between sets defined as,

$$h(i,j) = |V_i \cup V_j| - |V_i \cap V_j| = |V_i| + |V_j| - 2|V_i \cap V_j| = b + c, \tag{7.3}$$

is somewhat better. It is 0 when $V_i = V_j$, but its maximum, $|V_i \cup V_j|$, still depends on the set sizes. Returning to binary rows x_i and x_j of the data table corresponding to entities i and j, it is easy to see that distance $h(i,j)$ indeed equals the squared Euclidean, and indeed city-block, distance between x_i and x_j:

$$h(i,j) = \sum_{v \in V} (x_{iv} - x_{jv})^2 = \sum_{v \in V} |x_{iv} - x_{jv}|.$$

The latter equation follows from the fact that $(x - x')^2 = |x - x'|$ for any binary x and x'. Measure $h(i,j)$ (7.3) is referred to as Hamming distance between the binary vectors.

The relative distance $h(i,j)/N$ is known as the mismatch coefficient between sets:

$$m(i,j) = h(i,j)/N = (b + c)/N \tag{7.4}$$

Its complement to unity, the so-called match coefficient,

$$s(i,j) = (N - h(i,j))/N = (a + d)/N, \tag{7.5}$$

is popular as well.

Index $s(i,j)$ is always between 0 and 1. It is 1 when $V_i = V_j$ and it is 0 when V_i and V_j are complementary parts of I.

An issue with $s(i,j)$ as a similarity measure is that it depends on the size N of I which may be irrelevant, especially in comparing two "little things in the big world" such as two text documents among thousands of others.

The so-called Jaccard coefficient [57],

$$J(i,j) = \frac{|V_i \cap V_j|}{|V_i \cup V_j|} = \frac{a}{a + b + c}, \tag{7.6}$$

does not depend on N. (It is sometimes referred to as the Tanimoto coefficient.) This measure takes into account only features that are present at either i or j or both and handles them in such a way that the value of J does not depend on the total feature set. It ranges between 0 and 1, and it is 1 when $V_i = V_j$

and 0 when V_i and V_j are disjoint. It is good mathematically, too, because its complement to unity, the dissimilarity $d_J = 1 - J$, satisfies the so-called triangle inequality, $d_J(i,j) \leq d_J(i,l) + d_J(j,l)$ for any $i, j, l \in I$ (for a proof see [90], p. 175). A popular algorithm for clustering categorical data, ROCK [43], uses the Jaccard coefficient.

However, there appears to be an intrinsic flaw in the Jaccard coefficient in that it systematically underestimates the similarity. To demonstrate this, let us consider two typical situations [96].

1. **Undervalued overlap.** When the sizes of sets V_i and V_j are about the same and their overlap is about half of the elements in each of them, the Jaccard coefficient is about 1/3, while one would expect the similarity score to be, in this case, about 1/2. To make the coefficient equal to 1/2, the overlap must contain about 2/3 of the elements from each of the sets, which intuitively should correspond to a score of 2/3.

2. **Undervalued part**. When V_i is part of V_j being smaller than V_j in size, then J is just equal to the proportion of V_i in V_j. If, for instance, the size of V_i is 0.2 of the size of V_j, then the value of J also will be 0.2. Such a small value contradicts our intuition on the relationship between entities i and j because the fact that $V_i \subset V_j$ may intuitively mean that they may be highly related, semantically in the case of text documents or evolutionarily in the case of genomes.

Is there any remedy that can be suggested? Yes. One can note that in the former example, each of the ratios $|V_i \cap V_j|/|V_i|$ and $|V_i \cap V_j|/|V_j|$ is equal to one half. In the latter example, one of them is still 0.2 but the other is 1! Thus, an appropriate similarity index should combine these two ratios. Practitioners usually take the maximum or minimum of them by relating $|V_i \cap V_j|$ to $\min(|V_i|, |V_j|)$ or $\max(|V_i|, |V_j|)$, respectively. Also, the geometric mean,

$$g(i,j) = \frac{|V_i \cap V_j|}{\sqrt{|V_i||V_j|}} = \frac{a}{\sqrt{(a+b)(a+c)}}, \tag{7.7}$$

or arithmetic mean,

$$f(i,j) = \left(\frac{|V_i \cap V_j|}{|V_i|} + \frac{|V_i \cap V_j|}{|V_j|}\right)/2 = a\frac{2a+b+c}{(a+b)(a+c)}, \tag{7.8}$$

can be utilized. These two indexes have all the advantages of the Jaccard coefficient and, in fact, are co-monotone with it, but they evaluate similarity in the two special cases above in a way that is closer to our intuition. In the case of two equal size sets overlapping by half, each g and f is one half too. In the case when one set is a 20% subset of the other, $g = 0.45$ and $f = 0.60$.

Distance and inner product

As mentioned above, the distance d_{ij} between rows x_i and x_j representing $i, j \in I$ in the original binary data table is precisely the Hamming distance $h(i,j)$ (7.3). The inner product of the two binary rows is the overlap (7.2). To get more subtle measures one needs to employ subtler tools. In particular, the normalized product, $(x_i, x_j)/\sqrt{(x_i, x_i)(x_j, x_j)}$ yields the geometric mean $g(i,j)$ (7.7).

When the feature means have been subtracted according to recommendations of the data recovery approach, the inner product of rows in the pre-standardized data matrix leads to an interesting similarity index [96]. With $a_v = p_v$, the proportion of entities in which feature v is present, and $b_v = 1$, the range, the scalar product of the standardized rows has been derived in (5.42) on page 163 as,

$$a_{ij} = \sum_{v \in V} y_{iv} y_{jv} = |V_i \cap V_j| - |V_i| - |V_j| + t(i) + t(j) + \gamma, \tag{7.9}$$

where γ is a constant and $t(i)$ (or $t(j)$) is the total frequency weight of features that are present at i (or, at j). (The more frequent is the feature, the less its frequency weight.)

Similarity index (7.9) is a linear analogue to the arithmetic mean coefficient t (7.8) and, thus, has properties similar to those of t. However, it is further adjusted according to the information content of features in h and g, which is evaluated over all entities by counting their frequencies. Also, it can be either positive or negative, thus introducing the expression power of the sign into similarity measurement, and should be further explored.

7.3.2 Dis/similarity between partitions

Similarity between partitions is measured based on their cross classification, that is, the contingency table presented in Table 7.6. Rows $k = 1, ..., K$ correspond to clusters in partition $S = \{S_1, ..., S_K\}$ and columns $l = 1, ..., L$ to clusters in partition $T = \{T_1, ..., T_L\}$, and the (k,l)-th entry, to the number of co-occurrences of S_k and T_l, denoted by N_{kl}. That means $N_{kl} = R_k \cap S_l$. Summary frequencies of row and column categories, N_{k+} and N_{+l} respectively, are presented in Table 7.6 on the margins titled "Total."

In clustering, comparing partitions is traditionally done via representing them by clique graphs or, almost equivalently, by binary relation matrices. Given a partition $S = \{S_1, ..., S_K\}$ on I, its clique graph Γ_S has I as its vertice set; edges connect any two entities that are in the same cluster $i, j \in S_k$ for some $k = 1, ..., K$. The matrix r_S is defined so that $r_{ij} = 1$ for any i, j belonging to the same class S_k for some $k = 1, ..., K$; otherwise, $r_{ij} = 0$. To correspond to graph Γ_S exactly, the matrix must be modified by removing the diagonal

Table 7.6: A contingency table or cross classification of two partitions, $S = \{S_1, ..., S_K\}$ and $T = \{T_1, ..., T_L\}$ on the entity set I.

Cluster	1	2	...	L	Total
1	N_{11}	N_{12}	...	N_{1L}	N_{1+}
2	N_{21}	N_{22}	...	N_{2L}	N_{2+}
...	...	...	....	...	...
K	N_{K1}	N_{K2}	...	N_{KL}	N_{K+}
Total	N_{+1}	N_{+2}	...	N_{+L}	N

and one of the halves separated by the diagonal, either that under or above the diagonal, because of their symmetry.

Let us consider graph Γ_S as the set of its edges. Then, obviously, the cardinality of Γ_S is $|\Gamma_S| = \sum_{k=1}^{K} \binom{N_{k+}}{2}$ where $\binom{N_{k+}}{2} = N_{k+}(N_{k+} - 1)/2$ is the number of edges in the clique of Γ_S corresponding to cluster S_k. In terms of the corresponding binary matrices, this would be the number of $r_{ij} = 1$ for $i, j \in S_k$, that is, N_k^2 standing for $\binom{N_{k+}}{2}$ here. With a little arithmetic, this can be transformed to:

$$|\Gamma_S| = (\sum_{k=1}^{K} N_{k+}^2 - N)/2 \tag{7.10}$$

To compare graphs Γ_S and Γ_T as edge sets one can invoke the four-fold table utilized in section 7.3.1 for comparing sets (see Table 7.5). Table 7.7 presents it in the current context [2].

Table 7.7: Four-fold table for partition graphs.

Graph	Γ_T	$\bar{\Gamma}_T$	Total
Γ_S	a	b	$\|\Gamma_S\|$
$\bar{\Gamma}_S$	c	d	c+d
Total	$\|\Gamma_T\|$	b+d	$a+b+c+d = \binom{N}{2}$

The elements of Table 7.7 are the cardinalities of the edge sets and their intersections. The only cardinalities of interest in what follows are $|\Gamma_S|$, $|\Gamma_T|$ and a. The first is presented in (7.10) and a similar formula holds for the second, $|\Gamma_T| = (\sum_{l=1}^{L} N_{+l}^2 - N)/2$. The third, a, is similar too because it is the cardinality of the intersection of graphs Γ_S and Γ_T, which is itself a clique graph corresponding to the intersection of partitions S and T. Thus,

$$a = (\sum_{k=1}^{K}\sum_{l=1}^{L} N_{kl}^2 - N)/2 \tag{7.11}$$

which is a partition analogue to the overlap set similarity measure (7.2). It is of interest because it is proportional (up to the subtracted N which is due to the specifics of graphs) to the inner product of matrices r_S and r_T, which is frequently used as a measure of similarity between partitions on its own.

Among other popular measures of proximity between partitions are partition graph analogues to the mismatch and match coefficients m and s, (7.4) and (7.5), Jaccard coefficient J (7.6) and geometric mean g (7.7). These analogues are referred to as distance (mismatch), Rand, Jaccard and Fowlkes-Mallows coefficients, respectively, and can be presented with the following formulas:

$$M = (|\Gamma_S| + |\Gamma_T| - 2a)/\binom{N}{2} \tag{7.12}$$

$$Rand = 1 - (|\Gamma_S| + |\Gamma_T| - 2a)/\binom{N}{2} = 1 - M \tag{7.13}$$

$$J = \frac{a}{|\Gamma_S| + |\Gamma_T| - a} \tag{7.14}$$

and

$$FM = \frac{a}{\sqrt{|\Gamma_S||\Gamma_T|}}. \tag{7.15}$$

[It should be noted that in the original formulation of FM by Fowlkes and Mallow [33] no N is subtracted from the sum of squares in $|\Gamma_S|$ and $|\Gamma_T|$. The Rand coefficient [112] is frequently used in the standardized form proposed in [56].] Also, FM was observed to work better than the other two in experiments conducted in [19]. This should be of no wonder because its subset-to-subset counterpart, the geometric mean (7.7), is an obvious improvement over the others' subset-to-subset counterparts, (7.3) and (7.6).

It should be noted also that those partition-to-partition coefficients that are linear with respect to elements of the four-fold table are equal, up to the constant N, to the corresponding between-set measures averaged according to the bivariate distribution of the cross classification of S and T. This is true for the intersection a (7.11), distance M (7.12) and *Rand* (7.13).

In particular, intersection a (7.11) averages the values N_{kl} themselves. Distance M (7.12) and *Rand* (7.13) average values $N_{k+} + N_{+l} - 2N_{kl}$ and

Table 7.8: Model confusion data.

Attribute	T	$\bar{T}$	Total
S	$(1-\epsilon)n$	ϵn	n
$\bar{S}$	ϵn	$(1-\epsilon)n$	n
Total	n	n	$2n$

$N - N_{k+} - N_{+l} + 2N_{kl}$ respectively, as related to the set-theoretic difference between S_k and T_l. This suggests a potential extension of the partition similarity indexes by averaging other, nonlinear, measures of set similarity such as the means (7.7) and (7.8) or even the Quetelet coefficients. The averaged Quetelet coefficients, G^2 and Q^2, are akin to traditional contingency measures, which are traditionally labeled as deliberately attached to the case of statistical independence. However, the material of sections 2.2.3 and 3.4.4 demonstrates that the measures can be utilized just as summary association measures in their own right with no relation to the statistical independence framework.

Example 7.56. Distance and chi-squared according to a model confusion data

Consider the following contingency table (Table 7.8) that expresses the idea that binary features S and T significantly overlap and differ by only a small proportion ϵ of the total contents of their classes. If, for instance, $\epsilon = 0.1$, then the overlap is 90%; the smaller the ϵ the greater the overlap.

Let us analyze how this translates in values of the contingency coefficients described above, in particular, the distance and chi-squared. The distance, or mismatch coefficient, is defined by formulas (7.12), (7.11) and 7.10) so that it equals the sum of the marginal frequencies squared minus the doubled sum of squares of the contingency elements. To normalize, for the sake of convenience, we divide all items by the total $4n^2$, which is $2n$ squared, rather than by the binomial coefficient $\binom{2n}{2}$. This leads to

$$M = (4n^2 - 2(2(1-\epsilon)^2 + 2\epsilon^2)n^2)/(4n^2) = 2\epsilon(1-\epsilon).$$

To the calculate chi-squared, we use formula (2.12) from section 2.2.3. According to this formula, X^2 is the total of contingency entries squared and related to both column and row marginal frequencies minus one:

$$X^2 = 2(1-\epsilon)^2 + 2\epsilon^2 - 1 = 1 - 4\epsilon(1-\epsilon).$$

These formulas imply that for these model data, $X^2 = 1 - 2M$.

For example, at $\epsilon = 0.05$ and $\epsilon = 0.1$, $M = 0.095$ and $M = 0.18$ respectively. The respective values of X^2 are 0.64 and 0.81.

□

Matching-based similarity versus Quetelet association

We can distinguish between two types of subset-to-subset and respective partition-to-partition association measures:

M: matching between subsets/partitions measured by their overlap a in (7.2), (7.11) and (7.13), and

C: conditional probability and Quetelet indexes such as (2.10) and (2.12) and (2.13).

Considered as they are, in terms of co-occurrences, these two types are so different that, to the author's knowledge, they have never been considered simultaneously. There is an opinion that, in the subset-to-subset setting, the former type applies to measuring similarity between entities whereas the latter type to measuring association between features.

However, there exists a framework, that of data recovery models for entity-to-entity similarity matrices, in which these two types represent the same measure of correlation applied in two slightly different settings. In this framework, each partition $S = \{S_1, ..., S_K\}$ of the entity set I, or a nominal variable whose categories correspond to classes S_k $(k = 1, ..., K)$, can be represented by a similarity matrix $s = (s_{ij})$ between $i, j \in I$ where $s_{ij} = 0$ for i and j from different classes and a positive real when i and j are from the same class. Consider two definitions:

M: $s_{ij} = 1$ if $i, j \in S_k$ for some $k = 1, ..., K$;

C: $s_{ij} = 1/N_k$ if $i, j \in S_k$ for some $k = 1, ..., K$ where N_k is the number of entities in S_k.

The assumption is that all similarities within S are the same, in the case M, or inversely proportional to the class size so that the larger the class the smaller the similarity, in the case C. The latter reflects the convention that the more frequent is a phenomenon the less information value in it.

Matrix $s = (s_{ij})$ corresponding to a partition S has dimension $N \times N$. Its average $\bar{s}$ is $\bar{s} = G(S) = 1 - \sum_k p_k^2$, Gini index, in the case M, or $\bar{s} = 1/N$, in the case C.

Let now S and T be two partitions on I. Corresponding similarity matrices $s = (s_{ij})$ and $t = (t_{ij})$, as well as their centered versions $s - \bar{s}$ and $t - \bar{t}$, can be considered $N \times N$ vectors. Then the inner product of these vectors, in the case M, is equal to $\sum_{k=1}^{K} \sum_{l=1}^{L} N_{kl}^2$ featured in equation (7.11) whereas the inner product of their centered versions in the case C is the Pearson chi-squared. This shows that the two types of partition-to-partition measures can be related to the two types of entity-to-entity similarities, in the data recovery framework.

Dissimilarity of a set of partitions

In some situations, especially when the same clustering algorithm has been applied many times at different initial settings or subsamples, there can be many partitions to compare.

A good many-to-many dis/similarity measure can be produced by averaging a pair-wise dis/similarity between partitions. Let us denote given partitions on I by S^1, S^2,..., S^m where $m > 2$. Then the average distance $M(\{S^t\})$ will be defined by formula:

$$M(\{S^t\}) = \frac{1}{m^2} \sum_{u,w=1}^{m} M(S^u, S^w) \tag{7.16}$$

where $M(S^u, S^w)$ is defined by formula (7.12).

An interesting measure was suggested in [101] based on the average partition matrix which is an entity-to-entity similarity matrix defined by

$$\mu(i,j) = \frac{\sum_{t=1}^{m} s_{ij}^t}{m^*(i,j)} \tag{7.17}$$

where s^t is the binary relation matrix corresponding to S_t with $s_{ij}^t = 1$ when i and j belong to the same class of S^t and $s_{ij}^t = 0$, otherwise, with $m^*(i,j)$ denoting the number of partitions S^t at which both i and j are present. The latter denotation concerns the case when (some of) partitions S^t have been found not necessarily at the entire entity set I but at its sampled parts.

In the situation in which all m partitions coincide, all values $\mu(i,j)$ are binary being either 1 or 0 depending on whether i and j belong to the same class or not, which means that the distribution of values $\mu(i,j)$ is bimodal in this case. The further away partitions from the coincidence, the further away the distribution of $\mu(i,j)$ from the bimodal. Thus, the authors of [101] suggest watching for the distribution of $\mu(i,j)$, its shape and the area under the empirical cumulative distribution,

$$A(\{S^t\}) = \sum_{l=2}^{L} (\mu_l - \mu_{l-1}) F(\mu_l) \tag{7.18}$$

where $\mu_1, \mu_2, ..., \mu_L$ is the sorted set of different entries $\mu(i,j)$ in the average matrix and $F(\mu)$ is the proportion of entries $\mu(i,j)$ that are smaller than μ.

In fact, the average distance (7.16) also characterizes the average partition matrix $\mu(i,j)$. To see that, let us assume, for the sake of simplicity, that all partitions S^t are defined on the entire set I and denote the average value in the matrix by $\bar{\mu}$ and the variance by σ_μ^2.

Statement 7.20. *The average distance $M(\{S^t\})$ (7.16) is proportional to $\bar{\mu}(1-\bar{\mu}) - \sigma_\mu^2$.*

Proof: According to definition, $M(\{S^t\}) = \sum_{u,w=1}^{m} \sum_{i,j\in I}(s_{ij}^u - s_{ij}^w)^2/m^2$. This can be rewritten as $M(\{S^t\}) = \sum_{i,j\in I} \sum_{u,w=1}^{m}(s_{ij}^u + s_{ij}^w - 2s_{ij}^u s_{ij}^w)/m^2$. Applying the operations to individual items we obtain $M(\{S^t\}) = \sum_{i,j\in I}(\sum_{w=1}^{m} \mu(i,j)/m + \sum_{u=1}^{m} \mu(i,j)/m - 2\mu(i,j)\mu(i,j)) = 2\sum_{i,j\in I}(\mu(i,j) - \mu(i,j)^2)$. The proof of the statement follows then from the definition of the variance of the matrix, q.e.d.

As proven in statement 2.1., given a probability of smaller than the average values, the variance of a variable reaches its maximum when the variable is binary. The value $\bar{\mu}$ in this case can be assigned the meaning of being such a probability. Then the formula in statement 7.20. shows that the average distance M, in fact, measures the difference between the maximum and observed values of the variance of the average partition matrix. Either this difference or the variance itself can be used as an index of similarity between the observed distribution of values $\mu(i,j)$ and the ideal case at which all of them coincide.

7.4 Dealing with missing data

7.4.1 Imputation as part of pre-processing

Some data entries can be missing because of various reasons such as difficulties in obtaining data, or because of inaccuracy in data collecting or maintaining, or because of deficiencies in data producing devices. There are two approaches to dealing with missing data:

(1) **Within a method**: missing data are treated within a specially modified method for data analysis. This approach is popular with such techniques as regression analysis, but not with clustering.

(2) **Before a method**: imputation of missing entries is conducted as part of data pre-processing before any more specific analyses. Researchers do recognize that patterns of missings may relate to the data contents: in gene expression analyses, weak gene expression is more likely to go unnoticed, and in income surveys, respondents are likely to leave sensitive questions unanswered. Still, most research is being done under the assumption that missings emerge at random entries. In the remainder of this section, we refer to this approach only.

Among the methods of imputation of incomplete data, one can distinguish the following three approaches:

1. Conditional mean, in which every missing entry is substituted by a predicted value such as the variable's mean or the value predicted by a regression function or decision tree [73, 79, 111].

2. Maximum likelihood based methods, in which missings are imputed according to an estimated data density function [79, 118]

3. Least-squares approximation, in which the data are approximated with a low-rank data matrix [34, 65, 90].

Let us briefly discuss them in turn.

7.4.2 Conditional mean

Arguably the most popular method of imputation, at least among practitioners, is the substitution of a missing entry by the corresponding variable's mean, which will be referred to as the Mean algorithm. More subtle approaches use regression based imputation in which a regression-predicted value is imputed [73, 79]. Decision trees are used for handling missing categorical data in [111]. The Mean algorithm has been combined with nearest-neighbor based techniques in [130].

Two common features of the conditional mean approaches are: (1) missing entries are dealt with sequentially, one-by-one, and (2) most of them rely on a limited number of variables. The other approaches handle all missings simultaneously by taking advantage of using all the available data entries.

7.4.3 Maximum likelihood

The maximum likelihood approach relies on a parametric model of data generation, typically, the multivariate Gaussian mixture model. The maximum likelihood method is applied for both fitting the model and imputation of the missing data. The most popular is the so-called expectation-maximization (EM) algorithm [118] which exploits the popular idea of alternating optimization to maximize the maximum likelihood criterion as described in section 6.1.5. The popularity of the maximum likelihood approach is based on the fact that it is grounded on a precise statistical model. However, methods within this approach may involve unsubstantiated hypotheses and be computationally intensive.

7.4.4 Least-squares approximation

This is a nonparametric approach based on approximation of the available data with a low-rank bilinear factorial model akin to the principal component analysis model (5.4).

Methods within this approach, typically, work sequentially by producing one factor at a time to minimize the sum of squared differences between the available data entries and those reconstructed via the low-rank model. Two ways to implement this approach can be distinguished:

1. **Iterative least squares (ILS):** Fit a low-rank approximate data model (5.4) by using nonmissing entries only and then interpolate the missing values with values found according to the model [34, 90].

2. **Iterative majorized least squares (IMLS):** Start by filling in all missing entries with some value such as zero, then iteratively approximate thus completed data by updating the imputed values with those implied by a low-rank SVD based approximation [65].

These can be combined with nearest-neigbor based techniques as follows: take a row that contains a missing entry as the target entity x_i, find its K nearest neighbors in X, and form a matrix X_i consisting of the target entity and its neighbors. Then apply an imputation algorithm to the matrix X_i, imputing missing entries at the target entity only. Repeat this until all missing entries are filled in. Then output the thus completed data matrix.

A global-local approach proposed in [136] involves two stages. First stage: Use a global imputation technique to fill in all missings in the original matrix X so that entity-to-entity distances can be calculated with no concessions to the presence of missing values. Let us denote the resulting matrix X^*. Second stage: Apply a nearest-neighbor based technique to fill in the missings in X again, but, this time, based on distances computed with the completed data matrix X^*. This global-local approach involving IMLS on both of the stages, in the beginning with $m = 4$ and in the end with $m = 1$, is referred to as algorithm INI in [136]. In experiments reported in [136], nearest-neighbor based least squares imputation algorithms give similar results outperforming other least-squares algorithms including the Mean and nearest-neighbor-Mean imputation. When the proportion of random missings grows to 10% and, moreover, 25%, INI becomes the only winner [136].

Overall, the subject is in early stages of development. Potential mechanisms of missings are not well defined yet.

7.5 Validity and reliability

7.5.1 Index based validation

After applying a clustering algorithm, the user would like to know whether the results indeed reflect an innate property of the data or they are just an artifact generated by the algorithm. In other words, the user wants to know whether the cluster structure found is valid or not. It is not a difficult question when there is a probabilistic mechanism for data generation, which is known to the user. Unfortunately, this is not the typical case in cluster analysis. Typically, the data come from a source that cannot be modelled this way if it can be at all. There is one more cause of uncertainty here: the user's wish to have an aggregate representation of the data, at a level of resolution, which cannot be straightforwardly formalized and, moreover, depends on the task at hand.

There are two types of indexes used to quantify the validity: internal and external.

Internal indexes

An internal validity index in clustering is a measure of correspondence between a cluster structure and the data from which it has been generated. The better the index value, the more reliable the cluster structure. We give just a few formulations.

1. Measures of cluster *cohesion* versus *isolation*

1.1. Silhouette width.

The silhouette width of an entity $j \in I$ [62] is a popular measure defined as:

$$sil(j) = (b(j) - a(j))/max(a(j), b(j))$$

where $a(j)$ is the average dissimilarity of j with its cluster and $b(j)$ the smallest average dissimilarity of j from the other clusters. Values $a(j)$ and $b(j)$ measure cohesion and isolation, respectively. Entities with large silhouette width are well clustered while those with small width can be considered intermediate.

The greater the average silhouette width, the better the clustering. The measure has shown good results in some experiments [109].

1.2. Point-biserial correlation.

The point-biserial correlation between a partition and distances is a global measure that has shown very good results in experiments described in [84]. As any other correlation coefficient, it can be introduced in the context of the data recovery approach similar to that of the linear regression in section 5.1.2. We follow the presentation in [90].

Let us denote the matrix of between-entity distances by $D = (d_{ij})$. This can be just an input dissimilarity matrix, but in our context, D is a matrix of squared Euclidean distances. For a partition $S = \{S_1, ..., S_K\}$ on I let us consider the corresponding "ideal" dissimilarity matrix $s = (s_{ij})$ where $s_{ij} = 0$ if i and j belong to the same cluster S_k for some $k = 1, ..., K$, and $s_{ij} = 1$ if i and j belong to different clusters. Both matrices are considered here at unordered pairs of different i and j; the number of these pairs is obviously $N_* = N(N-1)/2$. Consider the coefficient of correlation (2.6) between these matrices as N_*-dimensional vectors. With elementary transformations, it can be proven that the coefficient can be expressed as follows:

$$r(D, s) = \frac{(d_b - d_w)\sqrt{N_b N_w}}{N_* \sigma_D} \tag{7.19}$$

where d_b is the average between cluster dissimilarity, d_w the average within cluster dissimilarity, N_b the number of unordered pairs of entities from different clusters, and N_w the total number of unordered pairs of entities taken from the same cluster; σ_D is the standard deviation of distances from their grand mean.

This coefficient perhaps should be referred to as the uniform correlation coefficient because it is the coefficient of correlation in the problem of uniform partitioning [90], that is, finding a partition in which all within cluster distances are equal to the same number α and all between cluster distances are equal to the same number β, so that its distance matrix is equal to $\lambda s + \mu$ where $\mu = \alpha$, $\lambda = \beta - \alpha$ and $s = (s_{ij})$ is the dissimilarity matrix of partition S defined above. In the framework of the data recovery approach, the problem of uniform partitioning is the problem of approximating matrix D with an unknown matrix $\lambda s + \mu$. This is the problem of regression analysis of D over s with the added stance that s is also unknown. Obviously, the least squares solution to the uniform partitioning problem is that one that maximizes the correlation coefficient $r(D, s)$ (7.19).

It appears, the problem indirectly involves the requirement that the cluster sizes should be balanced, which works well when the underlying clusters are of more or less similar sizes [90]. The criterion may fail, however, when the underlying clusters drastically differ in sizes. Good results shown by criterion (7.19) and its versions in experiments conducted by G. Milligan [84] may have been implied by the fact that cardinalities of clusters generated in these experiments were much similar to each other.

2. Indexes derived using the data recovery approach.

2.1. Attraction.

This is a data-recovery based analogue to the concept of silhouette width above, described in section 5.4.1. The attraction of entity $j \in I$ to its cluster is defined as $\beta(j) = a(j) - a/2$ where $a(j)$ is the average similarity of j to entities in its cluster and a the average within cluster similarity. If the data are given in feature-based format, the similarity is measured by the inner product of corresponding row vectors. Otherwise, it is taken from data as described in section 5.5.5.

Attraction indexes are positive in almost all clusterings obtained with the K-Means and Ward-like algorithms; still, the greater they are, the better the clustering. The average attraction coefficient can be used as a measure of cluster tightness.

2.2. Contribution of the cluster structure to the data scatter.

A foremost index of validity of a cluster structure in the data recovery approach is the measure of similarity between the observed data and that arising from a clustering model. Criterion $B(S, c)$, the cluster structure's contribution to the data scatter according to decomposition (5.13) in section 5.2, measures exactly this type of similarity: the greater it is, the better the partition fits into the data.

It should be pointed out that the data recovery based criteria are formulated in such a way that they seem to score cluster cohesion only. However, they implicitly do take into account cluster isolation as well, as proven in Chapter 5.

3. Indexes derived from probabilistic clustering models.

A number of indexes have been suggested based on probabilistic models of cluster structure; for a review, see [8, 39]. Results of using these indexes highly depend on the accepted model which, typically, the user has no possibility of verifying.

Use of internal indexes to estimate the number of clusters

The issue of how to determine the number of clusters in K-Means and other partitional algorithms has attracted a great deal of attention in the literature. As explained in section 1.2.6, this issue is not always meaningful. However, this does not mean that a review of the many attempts should be glossed over.

On the first glance, any validity measure, such as the average silhouette width and uniform correlation explained above, can be used for selecting the number of clusters: just do clustering for a range of values of K and select that one which makes the measure maximum (or minimum if the better fit corresponds to its decrease).

This idea was explored by many authors with respect to the within-cluster variance, W_K, the value of $W(S, c)$ at the found K-cluster partition. The index as is obviously cannot be used because it monotonely decreases when K grows: the greater is the number of clusters, the better fit. But its change, the first (or even second) difference

$$(W_K - W_{K+1})/T,$$

where T is the data scatter, can be considered a good signal: the number of clusters should stop rising when this difference becomes small. Hartigan [48] is credited with formally shaping a similar idea: take the index,

$$h_K = (W_K/W_{K+1} - 1)(N - K - 1),$$

and one by one increase K starting from 1; stop calculations when h_K becomes less than 10. An index by Calinski and Harabasz,

$$c_K = [(T - W_K)/(K-1)]/[W_K/(N-K)],$$

associated with Fisher distribution in statistics, showed a good performance in some experiments [86].

There are measures that compare W_K with its expected value under a uniform distribution. In particular, Krzanowski and Lai [71] showed that the expected value of W_K is proportional to $K^{-2/M}$ where M is the dimension of the space. They used ratio,

$$K^{2/M} W_K,$$

and its differences to derive the best number of clusters. Similar indexes have been proposed in the so-called Gap-statistic [129] and Jump-statistic [128].

The latter seems especially suitable: calculate the average 'distortion' of an axis associated with fitting K centroids to data with the K-Means algorithm, $w_K = W_K/M$ where M is the number of features and W_K the value of K-Means square error criterion at a partition found with K-Means, and calculate 'jumps' $j_K = w_K^{-M/2} - w_{K-1}^{-M/2}$. With $w_0 = 0$ and $w_1 = T/M$ where T is the data scatter, jumps can be defined for all reasonable values of $K = 0, 1, ...$ The maximum jump corresponds to the right number of clusters. This is supported with a mathematical derivation stating that if the data can be considered a standard sample from a mixture of Gaussian distributions and distances between centroids are great enough, then the maximum jump would indeed occur at K equal to the number of Gaussian components in the mixture [128].

Reviews of these and other indexes of the best number of clusters can be found in [22], [138], [129].

Experiments conducted by authors of [138] and others show that none of them can be considered a universal criterion.

In fact, the straightforward option does not necessarily work even for the average silhouette width and the uniform correlation. In particular, [109] reports that somewhat better results could be obtained by using the average silhouette width indirectly: for determining that no further splits of clusters are needed: the smaller the average silhouette width within a split cluster, the more likely it should not be split at all.

There is no universal criterion because, apart from obvious cases such as clearly seen differences between rock and water, light and dark, horse and birch tree, clusters are not only in the data but also in the mind of the user. The granularity of the user's vision and between-cluster boundaries, when no natural boundaries can be seen, should be left to the user. This can be formalized by introduction of relevant "external" criteria.

External indexes

The so-called external indexes compare a clustering found in data with another clustering either given by an expert or following from a knowledge domain or found by another clustering algorithm. Typically clusterings are sets or partitions of the entity set I. Thus, the external indexes are those described in sections 2.2.3 and 7.3.

7.5.2 Resampling for validation and selection

Resampling is a procedure oriented at testing and improving reliability of data based estimators and rules with respect to changes in data. Basically, it produces an empirical distribution of clustering results and related indexes to see the extent of dispersion of these with regard to data changes. This type of analysis has become feasible with the development of computer hardware and is in its infancy as a discipline. After having read through a number of recent publications involving resampling and done some computational experiments, the author can propose the following systematic of resampling in clustering.

Resampling involves the following steps:

A: Generation of a number of data sets, *copies*, related to the data under consideration.

B: Running an accepted algorithm (such as regression, classification tree, or clustering) independently at each of the copies.

C: Evaluating results.

D: Aggregating.

Let us describe these steps in greater detail bearing in mind that the data set under analysis has the standard entity-to-feature format:

A **Generation of copies** can be done by either *subsampling* or *splitting* or *perturbing* the data set:

A1 Subsampling: a proportion α, $0 < \alpha < 1$, is specified and αN entities are selected randomly without replacement in a subsample; the copy data is constituted by the rows corresponding to selected entities.

A2 Splitting: set I is randomly split in two parts of pre-specified sizes: the training and testing parts. This can be done as is, by generating a copy with a *train/testing split*, or via the so-called *cross-validation* at which the entity set is randomly split in a prespecified number of parts Q of approximately equal size, and Q copies are created simultaneously. A copy is created by considering one of the Q parts as the testing set, and its complement as the training set. Typically, Q is taken as either 2, 5 or 10. However, the case when Q is equal to the number of entities is also popular; it is called *leave-one-out* cross validation [50].

A3 Perturbing: the size of a copy here is the same as in the original data set but the entries are perturbed by:

A3.1 *Bootstrapping.* In this process, the data set is considered as that consisting of N rows. Uniform random sampling with replacement is applied N times to randomly pick up N row indexes, which are not necessarily different because of the replacements. A copy set is created then by combining the N selected rows into a data table. Obviously, some rows may be left out of the created data table since some rows may occur there two or more times. It is not difficult to estimate the probability of a row not being selected. At each action of picking out a row, the probability of a row not being selected is $1 - 1/N$, which implies that the approximate proportion of rows never selected in the sampling process is $(1 - 1/N)^N \approx 1/e = 36.8\%$ of the total number of entities. That means that any bootstrap copy on average includes only about 63.2% of the entities, some of them several times. A similar process can be applied to columns of the data matrix [30].

A3.2 *Adding noise.* In this process, a random real number is added to each entry. The added number is typically generated from the normal distribution with zero mean and a relatively small variance. In [64] a combination of bootstrapping and adding noise is described. For a K-Means found cluster structure (S, c), let us calculate the residuals e_{ik} in the data recovery model (5.11). A bootstrapping experiment is then defined by randomly sampling with replacement among the residuals e_{iv} and putting the sampled values in equations (5.11) to calculate the bootstrapped data.

B **Running an algorithm** on a copy is done the same way as on the original data set except for that if the copy is obtained by splitting, the algorithm is run over its training part only.

C **Evaluating results** depends on the copy type. If copies are obtained by subsampling or perturbing, the results on copies are compared to the results found with the original data set. If copies are obtained by splitting, the training part based results are evaluated on the test part. Let us describe this in more detail for each of the copy types.

C1 **Subsample case.** The cluster structure found from a subsample is compared with the cluster structure on the same subsample resulting from the application of the algorithm to the original data set. The comparison can be done with any index such as *Rand* (discussed in section 7.5.1) or chi-square contingency coefficients (discussed in section 2.2.3). For instance, in [78] the evaluation is done according to the averaged overlap index (7.11); in [104] relative distance between "sample" and "real" centroids is computed. This concerns situations in which subsample results do not depend on the subsample size, which is not so sometimes. For instance, if the fitted model involves a feature interval (like in APPCOD based descriptions), the interval's length on an αN subsample will be, on average, α times the interval's length on the original data. Then the interval must be stretched out in the proportion $1/\alpha$ [78].

C2 **Split case.** The evaluation of a model fitted to the training data is rather easy when the model is that of prediction or description as in the case of the regression or classification tree. In these, the training part fitted model is applied as is to the testing part and the error score of the prediction/description in the testing part is calculated. When the fitted model is a cluster structure such as produced by K-Means, its evaluation in the testing part is not straightforward and can be done differently. For instance, in [22] the following method, Clest, is proposed. First, apply K-Means to the testing parts anew. Then build a prediction model, such as a classification tree, on the training part and apply it to the testing part to predict, for each testing part entity, what cluster it should belong to. This also can be done by using the minimum distance rule applied to the testing parts at the training part based centroids [81]. In this way, two partitions are found on the testing part: that predicted from the training part and that found anew. The evaluation then is done by comparing these two partitions: the closer the better (see examples in section 3.3.2). An original use of the leave-one-out model for measuring similarity between entities is described in [52]; the authors use the minimum rather than average correlation over pair-wise correlations with one column removed; this also cleans the outlier effects.

C3 **Perturbation case.** The evaluation is especially simple when the data have been perturbed by adding noise, because the entities remain the same. Thus, the evaluation can be done by comparing the cluster structure found with the original data and cluster structure found at the perturbed data. In the case of partitions, any of the indexes of section 7.5.1 would fit. In [64] evaluation is done by testing the perturbed data partition against the original centroids; the error is proportional to the number of cases in which the minimum distance rule assigns an entity to a different cluster. In the case of bootstrap, evaluation can be done by comparing those parts of the two cluster structures, which relate to that aspect of data that was not changed by the sampling. In particular, Felsenstein [30] suggested bootstrapping columns (features) rather than entities in building hierarchical clusters and more generally phylogenetic trees. Evaluation of results is performed by comparing entity clusters found on the perturbed data with those found on the original data: the greater the number of common clusters, the better.

D **Aggregating results.** This can be done in various ways such as the following:

D1 **Combining evaluations.** Typically, this is just averaging the evaluations for each individual copy. The averaged score can be considered as a test result for the algorithm. In this way, one can select the best performing algorithm among those tested. This can also be applied to selection of parameters, such as the number of clusters, with the same algorithm [78], [22]. The results on perturbed copies can be used to score confidence in various elements of the cluster structure found at the original data set. For instance, in [30], each of the hierarchic clusters is accompanied by the proportion of copies on which the same algorithm produced the same cluster: the greater the proportion, the greater the confidence.

D2 **Averaging models.** Models found with different copies can be averaged if they are of the same format. The averaging is not necessarily done by merely averaging numerical values. For instance, a set of hierarchical clustering structures can be averaged in a structure that holds only those clusters that are found in a majority of the set structures [80]. In [19], centroids found at subsamples are considered a data set which is clustered on its own to produce the averaged centroids.

D3 **Combining models.** When models have different formats, as in the case of decision trees that may have different splits over different features at different copies, the models can be combined to form a 'committee' in such a way that it is their predictions rather than themselves that are averaged. Such is the procedure referred to as bagging in [50].

7.5.3 Model selection with resampling

Since a quality score such as the number of misclassified entities or the contribution to the data scatter may be different over different samples, resampling can be used not only for testing, but for learning, too. Of a set of algorithms one is selected at which an accepted quality score coefficient is optimized. This is called model selection. Examples of model selection can be found in Efron and Tibshirani [25] p. 243-247 (selection of decision tree sizes), Salzberg [117], p. 193-194 (averaging probabilities obtained with split-sample-based decision trees) and the following subsections.

Determining the number of clusters with K-Means

This subject has been treated in a number of publications [22, 78, 101, 129]. Let us describe a method proposed in [101]. This method exploits the properties of the average partition matrix $\mu(i,j)$ $(i,j \in I)$ and the empirical distribution of its entries described above on page 229.

Let us specify a number K, perform K-Means on many random subsamples from the entity set I, define the average partition matrix $\mu(i,j)$ based on partitions found and the area under its cumulative distribution, $A(K)$, as defined by formula (7.18). After having done this for $K = 2, 3, ...K_{\max}$, $B(K^{'})$ is defined as the maximum of $A(K)$, $K = 2, ..., K^{'}$. Then the relative area increase is expressed as $\Delta(K) = (A(K+1) - B(K))/B(K)$, $K = 3, 4, ...$ At $K = 2$, $\Delta(K)$ is defined as $\Delta(2) = A(2)$. Based on a number of simulation studies, the authors of [101] claim that the maximum of this index is indicative of the true number of clusters.

Model selection for cluster description

To learn a comprehensive description of a subset $S \subset I$, we apply a resampling process that can be considered a model of generalization in a learning device. According to this model, the learning process is organized in two stages. At the first stage, the learning device is presented with different samples of the data, in which S-entities are indicated. Based on these, the learning device develops a set of combined features that are most effective for separating S from the rest. At the second stage, the device is again presented with random supervised data samples. Now samples are used not for feature generation but rather for building an aggregate conjunctive description in terms of features generated at the first stage.

These stages can be characterized in brief as follows [95].

1. **Extracting features.** At this stage, a number of random samples from the pair (I, S) is generated, and the APPCOD algorithm of comprehensive description is applied to each of them. The best of these descriptions is selected to proceed. The quality of a description is evaluated by its error on the entire data set.

   To extend an α sample based description to the whole data set, each feature interval predicate is extended by stretching its interval out by the reverse coefficient $1/\alpha$. This generally breaks the property of APPCOD produced predicates to cover the entire within-S ranges of the features and thus restores the ability to have both false positive and false negative errors.

   To measure the error, one may use either the conventional total number of errors $FP + FN$ or the more subtle criterion $E = \beta FP/|I - S| + (1-\beta)FN/|S|$ of the weighted summary proportion of errors. Changing weight coefficient β allows one to take into account the relative importance of false positives versus false negatives. If S is a cancer type, false negatives should be avoided (β is to be close to 0); false positives are of an issue when S are loyal customers (β is to be close to 1). Otherwise, keep $\beta = 0.5$. Relative, not absolute, values of the errors appear in E to address the situations in which S is significantly smaller than $I - S$.

The features occurring in the best comprehensive description are considered effective for learning set S in I and added to the feature set.

2. **Selecting descriptions.** At this stage, a number of independent samples from (I, S) is generated again and an APPCOD produced description is found at each of them. However, this time the APPCOD is used as a feature selector, not extractor. Only features present after the first stage, not their combinations, are used in producing feature interval predicates. Then sample descriptions are aggregated according to a version of the majority rule: only features occurring at the majority of sample descriptions are selected. Their intervals are determined by averaging the left and right boundaries of intervals involved in the sample feature interval predicates.

The algorithm can be tuned up with two principal parameters: the maximum number of items permitted in a conjunctive description, l, and the number of arithmetic operations admitted in a combined feature, f. This proceeds in a loop over f and l, starting with a value of 1 for each and doing the loop over l within the loop over f. Based on our experimentation, these were set f=2 and the limits of l set from $l = 5(f - 1) + 1$ to $l = 5f$ by default. For any given pair, l and f, the generalization process applies a number of times (three, by default) and the best result is picked up and stored [95].

Example 7.57. Generalization with resampling at Body mass data.
Applied to the Body mass data in Table 1.7, this method produces either the body mass index Weight/(Height*Height) or the less expressive variables Weight/Height and Height - Weight depending on numbers of random samples at each step of the generalization process described above. When both numbers are small, the less expressive variables tend to appear. When the number of samples at the Aggregating step increases, the method produces the body mass index. This tendency is much less expressed when the number of samplings increases at the Generalization step. Curiously, these tendencies practically do not depend on the sample size. □

Example 7.58. Binary data: Republicans and democrats at the US congressional voting
Let us describe results of applying the APPCOD based model selection method on the data set of 1984 United States Congressional Voting Records from [113] which contains records of sixteen yes/no votings by 267 democrats and 168 republicans. It appears the set of republicans can be described by just one predicate pff/esa=1 admitting 11 FP and 0 FN, 2.53% of total error. Here pff and esa are abbreviations of issues "physician-fee-freeze" and "export-administration-act-south-africa," and 1 codes 'yes' and 0 'no'.

This means that all republicans, and only 11 democrats, voted consistently on both issues, either both yes or both no, which cannot be described that short without the operation of division. Another issue: the method describes each individual class asymmetrically, which allows us to check which of the classes, republican or democrat, is more coherent. Because of the asymmetry in APPCOD, one can safely claim that the class whose description has the minimum error is more coherent. Descriptions of democrats found at various subsamples always had 3-4 times more errors than descriptions of republicans, which fits very well into the popular images of these parties [95]. □

7.6 Overall assessment

The chapter reviews a number of issues of current interest in clustering as part of data mining.

First, the issue of finding a relevant feature space is considered. There are two approaches here: feature selection and feature transformation (extraction). Automatic transformation of the feature space is a major problem in data mining which is yet to be properly addressed. Arithmetically combining features within the context of a decision rule maker, specifically, the comprehensive description algorithm, can be of value in attacking the problem.

Second, issues and approaches in data pre-processing and standardization, including dealing with missing data, are discussed – advantages of the data recovery approach are pointed out.

Third, issues of cluster validity including those of determining the "right" number of clusters, are addressed in terms of: (a) internal and external indexes and (b) data resampling.

Internal indexes show the correspondence between data and clusters derived from them. Obviously, the best source of such indexes are the data recovery criteria and related coefficients. Unfortunately, none of the indexes proposed so far can be considered a universal tool because of both the diversity of data and cluster structures and the different levels of granulation required in different problems.

External indexes show correspondence between data-driven clusters and those known from external considerations. We provide a description of such indexes in section 7.3 in such a way that correspondences between set-to-set and partition-to-partition indexes are established, as well as relations between structural and association indexes.

Data resampling as a tool for model testing and selection is a relatively new addition. A systematic review of the approaches is given based on the most recent publications. A data resampling based model for learning a comprehensive description of a cluster is discussed.

Conclusion: Data Recovery Approach in Clustering

Traditionally clusters are built based on similarity. A found cluster of similar entities may be used as a whole to generalize and predict. This order of action is reversed in the data recovery approach. A property of similarity clusters, the possibility to aggregate data of individual entities into data of clusters, is taken here as the defining attribute. According to the data recovery approach, entities are brought together not because they are similar but because they can be used to better recover the data they have been built from. Is it not just a new name for old wine? Indeed, the closer the entities to each other the more each of them resembles the cluster's profile.

Yet the shift of the focus brings forward an important difference. In the data recovery approach, the concept of similarity loses its foundation status and becomes a derivative of the criterion of recovery. In conventional clustering, the emphasis is on presenting the user with a number of options for measuring similarity; the greater the choice the better. There is nothing wrong with this idea when the substantive area is well understood. But everything is wrong with this when the knowledge of the substantive area is poor. No user is capable of reasonably choosing a similarity measure in such a situation. A similarity measure should come from the data mining side. This is the case in which a data recovery approach can provide sound recommendations for the similarity measurement and, moreover, for the data pre-processing needed to balance items constituting the data recovery criterion.

The heart of the data recovery clustering framework is Pythagorean decomposition of the data scatter into two items, that explained by the cluster structure and that unexplained, the square error. The items are further decomposed in the contributions of individual entity–feature pairs or larger substructures. Some developments that follow from this:

1. **Pre-processing.**

   (a) The data scatter expresses the scattering of entities around the origin of the feature space, which thus must be put into a central, or normal, position among the entity points (this position is taken to be the grand mean).

(b) The data scatter is the sum of feature contributions that are proportional to their variances thus reflecting the distribution shapes; this allows for tuning feature normalization options by separating scale and shape related parts.

(c) The strategy of binary coding of the qualitative categories in order to simultaneously process them together with quantitative features, which cannot be justified in conventional frameworks, is supported in this framework with the following:

   i. Binary features appear to be the ultimate form of quantitative features, those maximally contributing to the data scatter.
   ii. The explained parts of category contributions sum up to association contingency coefficients that already have been heavily involved in data analysis and statistics, though from a very different perspective.
   iii. The association coefficients are related to the data normalization options, which can be utilized to facilitate the user's choice among the latter; this can be done now from either end, the process' input or output, or both.
   iv. The equivalent entity-to-entity similarity measure which has emerged in the data recovery context is akin to best heuristic similarity measures but goes even further by taking into account the information weights of categories.

2. **Clustering.**

   (a) Data recovery models for both K-Means and Ward clustering extend the Principal Component Analysis (PCA) model to the cases in which scoring vectors are to be compulsory binary or tertiary, respectively. This analogy should not be missed because the PCA itself is conventionally considered but a heuristic method for extracting the maximum variance from the data, which is not quite correct. In fact, the PCA can be justified by using a data recovery model, as shown in section 5.1.3, and then extended to clustering.

   (b) K-Means is the method of alternating minimization applied to the square error criterion.

   (c) Agglomerative and divisive Ward clustering involve somewhat "dual" decompositions of the data scatter. That for divisive clustering is a natural one, treating the scatter's unexplained part as a whole and further decomposing the explained part into items related to the contributions of individual clusters and features. In contrast, the decomposition for agglomerative clustering hides the structure of the explained part, which probably can explain why no specific tools for the interpretation of cluster hierarchies have been proposed before.

(d) By exploiting the additive structure of the data recovery models in the manner following that of the PCA method, one-by-one clustering methods are proposed to allow for effective computational schemes as well as greater flexibility in a controlled environment. In particular, the intelligent version of K-Means, iK-Means, can be used for incomplete clustering with removal of devious or, in contrast, overly normal items, if needed.

(e) Local search algorithms presented lead to provably tight clusters – the fact expressed with the attraction coefficient, a theory-based analogue to popular criteria such as the silhouette width coefficient.

(f) The approach is extended to contingency and flow data by taking into account the property that each entry is a part of the whole. The entries are naturally standardized into the Quetelet coefficients; the corresponding data scatter appears to be equal to the chi-squared contingency coefficient.

(g) The inner product can be used as an equivalent device in the correspondingly changed criteria, thus leading to similarity measures and clustering criteria – some of those are quite popular and some are new, still being similar to those in use.

3. **Interpretation aids.**

(a) The models provide for using the traditional cluster centroids as indicators of cluster tendencies. Also, more emphasis is put on the standardized, not original, values thus relating them to the overall norms (averages).

(b) Inner products of individual entities and centroids, not distances between them that are used conventionally, express entity contributions to clusters. This relates to the choice of a cluster representative as well: not by the distance but by the inner product.

(c) The decomposition of the data scatter over clusters and features in table ScaD provides for comparative analysis of the relative contributions of all elements of the cluster structure, especially with Quetelet coefficients (table QScaD), to reveal the greatest of them.

(d) APPCOD, a method for conceptual description of individual clusters based on the table ScaD, can serve as a supplementary or complementary tool to classical decision trees as interpretation and description aids.

(e) In hierarchical classification, the conventional decomposition of the data scatter over splits and clusters has been supplemented with similar decompositions of feature variances, covariances and individual entries; these may be used as aids in the interpretation of the tendencies of hierarchical clusters.

(f) The decomposition of the data scatter over an upper cluster hierarchy can now be visualized with the concept of the box-chart, extending the conventional pie-charts.

Much room remains for further developments and extensions of the data recovery approach.

First, data specifics, such as those that we considered here only for the cases of mixed and contingency data tables, should be taken into account. Consider, for instance, digitized image data. Here entities are pixels organized in a grid. Conventionally, image data are compressed and processed with techniques exploiting their spatial character with such constructions as quadtrees and wavelets, which are unrelated in the current thinking. The data recovery model with upper cluster hierarchies, used here for developing divisive Ward-like clustering, in fact can be considered as an extension of both quadtrees and wavelets. This may potentially lead to methods of image processing while simultaneously compressing them. Another promising area is applying the data recovery approach to analysis of temporal or spatio-temporal data. An advantage of modeling data with a cluster model is that the temporal trajectory of a cluster centroid can be modeled as a specific, say exponential, function of time. Gene expression data contain measurements of several properties made over the same gene array spots; they should be considered another promising direction.

Second, the models themselves can be much improved beyond the simplest formats employed in the book. These models, in fact, require any data entry to be equal to a corresponding entry in a centroid or a combination of centroids. This can be extended to include "gray" memberships, transformed feature spaces, and logically described clusters. The least squares criterion can be changed for criteria that are less sensitive to data variation. The least moduli criterion is a most obvious alternative. Possibilities of probabilistic modelling should not be discarded either. In the current setting, the Pythagorean decompositions much resemble those used for the analysis of variance of a single variable over various groupings. Related probabilistic models, included in most statistics texts, seem, however, overly rigid and restrictive. Hopefully, cluster analysis may provide a ground for seeking more relevant probabilistic frameworks.

Third, clustering methods can be extended from the purely local search techniques presented in the book. One of the directions is building better versions with provable approximation estimates. Another direction is applying the evolutionary and multi-agent approaches of computational intelligence.

Bibliography

[1] K. Ali and M. Pazzani (1995) Hydra-mm: Learning multiple descriptions to improve classification accuracy, *International Journal on Artificial Intelligence Tools, 4*, 115-133.

[2] P. Arabie, L. Hubert, and G. De Soete (Eds.) (1996) *Classification and Clustering*, Singapore: World Scientific.

[3] S. Baase (1991) *Computer Algorithms, Second Edition*, Reading, Ma: Addison-Wesley.

[4] Bailey, K.D. (1994) *Typologies and Taxonomies: An Introduction to Classification Techniques*, London: Sage Publications.

[5] Y. Barash and N. Friedman (2002) Context-specific Bayesian clustering for gene expression data, *Journal of Computational Biology*, 9, 169-191.

[6] J.P. Benzecri (1992) *Correspondence Analysis Handbook*, New York: Marcel Dekker.

[7] J.C. Bezdek, J.M. Keller, R. Krishnapuram, L.I. Kuncheva, and N.R. Pal (1999) Will the *real* Iris data please stand up?, *IEEE Transactions on Fuzzy Systems*, 7, no. 3, 368-369.

[8] H.H. Bock (1996) Probability models and hypothesis testing in partitioning cluster analysis. In P. Arabie, C. D. Carroll, and G. De Soete (Eds.) *Clustering and Classification*, River Edge, NJ: World Scientific Publishing, 377-453.

[9] H.H.Bock (1999) Clustering and neural network approaches. In W. Gaul and H. Locarek-Junge (Eds.) *Classification in the Information Age*, Berlin-Heidelberg: Springer, 42-57.

[10] Boley, D.L. (1998). Principal direction divisive partitioning. *Data Mining and Knowledge Discovery, 2(4)*, 325-344.

[11] L. Breiman, J.H. Friedman, R.A. Olshen, and C.J. Stone (1984) *Classification and Regression Trees*, Belmont, Ca: Wadswarth International Group.

[12] T.M. Cover and J.A. Thomas (1991) *Elements of Information Theory*, Wiley.

[13] L.L. Cavalli-Sforza (2001) *Genes, Peoples, and Languages*, London: Penguin Books.

[14] *Clementine 7.0 User's Guide Package* (2003) Chicago: SPSS Inc.

[15] W. W. Cohen (1995) Fast effective rule induction, In Armand Prieditis and Stuart Russell (Eds.) *Proc. of the 12th International Conference on Machine Learning*, 115-123, Tahoe City, CA, 1995. Morgan Kaufmann.

[16] S. Deerwester, S.T. Dumais, G.W. Furnas, T.K. Landauer, and R. Harshman (1990) Indexing by Latent Semantic Analysis, *Journal of the American Society for Information Science, 41*, 391-407.

[17] G. Der and B.S. Everitt (2001) *Handbook of Statistical Analyses Using SAS, Second Edition*, CRC Press.

[18] M. Devaney and A. Ram (1997) Efficient feature selection in conceptual clustering, In: *Proceedings of 14th International Conference on Machine Learning*, 92-97, Morgan Kaufmann.

[19] S. Dolnicar and F. Leisch (2000) Getting more out of binary data: segmenting markets by bagged clustering, Working paper no. 71, Vienna University of Economics and Business Administration, 22 p.

[20] S. Draghici (2003) *Data Analysis Tools for DNA Microarrays*, Boca Raton/London/New York: Chapman & Hall/CRC.

[21] R.O. Duda and P.E. Hart (1973) *Pattern Classification and Scene Analysis*, New York: J.Wiley & Sons.

[22] S. Dudoit and J. Fridlyand (2002) A prediction-based resampling method for estimating the number of clusters in a dataset, *Genome Biology, 3*, n. 7, 1-21.

[23] M.H. Dunham (2003) *Data Mining: Introductory and Advanced Topics*, Upper Saddle River, NJ: Pearson Education Inc.

[24] S. Dzeroski and N. Lavrac (Eds.) (2001) *Relational Data Mining*, Berlin: Springer-Verlag.

[25] B. Efron and R.J. Tibshirani (1993) *An Introduction to the Bootstrap*, Chapman and Hall.

[26] M. Ester, A. Frommelt, H.-P. Kriegel, and J. Sander (2000) Spatial data mining: Database primitives, algorithms and efficient dbms support, *Data Mining and Knowledge Discovery*, 4, 193-216.

[27] B.S. Everitt and G. Dunn (2001) *Applied Multivariate Data Analysis*, London: Arnold.

[28] B.S. Everitt, S. Landau, and M. Leese (2001) *Cluster Analysis (4th edition)*, London: Arnold.

[29] U.M. Fayyad, G. Piatetsky-Shapiro, P. Smyth, and R. Uthurusamy (Eds.) (1996) *Advances in Knowledge Discovery and Data Mining*, Menlo Park, Ca: AAAI Press/The MIT Press.

[30] J. Felsenstein (1985) Confidence limits on phylogenies: an approach using the bootstrap, *Evolution*, 39, 783-791.

[31] D.W. Fisher (1987) Knowledge acquisition via incremental conceptual clustering, *Machine Learning, 2*, 139-172.

[32] K. Florek, J. Lukaszewicz, H. Perkal, H. Steinhaus, and S. Zubrzycki (1951) Sur la liason et la division des points d'un ensemble fini, *Colloquium Mathematicum, 2*, 282-285.

[33] E.B. Fowlkes and C.L. Mallows (1983) A method for comparing two hierarchical clusterings, *Journal of American Statistical Association, 78*, 553-584.

[34] K.R Gabriel and S. Zamir (1979) Lower rank approximation of matrices by least squares with any choices of weights, *Technometrics, 21*, 489-298.

[35] A.P. Gasch and M.B. Eisen (2002) Exploring the conditional coregulation of yeast gene expression through fuzzy k-means clustering, *Genome Biology*, 3, 11.

[36] M. Girolami (2002) Mercer kernel based clustering in feature space, *IEEE Transactions on Neural Networks*, 13, 780-784.

[37] R. Gnanadesikan, J.R. Kettenring, and S.L. Tsao (1995) Weighting and selection of variables, *Journal of Classification*, 12, 113-136.

[38] G.H. Golub and C.F. Van Loan (1989) *Matrix Computations*, Baltimore: J. Hopkins University Press.

[39] A.D. Gordon (1999) *Classification (2nd edition)*, Boca Raton: Chapman and Hall/CRC.

[40] J.C. Gower (1967) A comparison of some methods of cluster analysis, *Biometrics, 23*, 623-637.

[41] J.C. Gower and G.J.S. Ross (1969) Minimum spanning trees and single linkage cluster analysis, *Applied Statistics, 18*, 54-64.

[42] S.B. Green and N.J. Salkind (2003) *Using SPSS for the Windows and Macintosh: Analyzing and Understanding Data (3rd Edition)*, Prentice Hall.

[43] S. Guha, R. Rastogi, and K. Shim (2000) ROCK: A robust clustering algorithm for categorical attributes, *Information Systems, 25*, n. 2, 345-366.

[44] J. Han and M. Kamber (2001)*Data Mining: Concepts and Techniques*, Morgan Kaufmann Publishers.

[45] P. Hansen and N. Mladenovic (2001) J-means: a new local search heuristic for minimum sum-of-squares clustering, *Pattern Recognition, 34*, 405-413.

[46] J.A. Hartigan (1967) Representation of similarity matrices by trees, *Journal of the American Statistical Association, 62*, 1140-1158.

[47] J.A. Hartigan (1972) Direct clustering of a data matrix, *Journal of the American Statistical Association, 67*, 123-129.

[48] J.A. Hartigan (1975) *Clustering Algorithms*, New York: J.Wiley & Sons.

[49] T. Hastie, R. Tibshirani, M.B. Eisen, A. Alizadeh, R. Levy, L. Staudt, W.C. Chan, D. Botstein, and P. Brown (2000) 'Gene shaving' as a method for identifying distinct sets of genes with similar expression patterns. *Genome Biology, 1*, http://genomebiology.com/ 2000/1/2/research/0003/abstract.

[50] T. Hastie, R. Tibshirani, and J.R. Fridman (2001) *The Elements of Statistical Learning*, New York: Springer.

[51] S. Haykin (1999) *Neural Networks, 2nd ed.*, New Jersey: Prentice Hall.

[52] L. J. Heier, S. Kruglyak, and S. Yooseph (1999) Exploring expression data: Identification and analysis of coexpressed genes, *Genome Research, 9*, 1106-1115.

[53] M.O. Hill (1979) TWINSPAN: a FORTRAN program for arranging multivariate data in an ordered two-way table by classification of the individuals and attributes. *Ecology and Systematics*, Ithaca, NY: Cornell University.

[54] N.S. Holter, M. Mitra, A. Maritan, M. Cieplak, J.R. Banavar, and N.V. Fedoroff (2000) Fundamental patterns underlying gene expression profiles: Simplicity from complexity, *Proceedings of the National Academy of Sciences of the USA, 97*, no. 15, 8409-8414.

[55] K.J. Holzinger and H.H. Harman (1941) *Factor Analysis*, Chicago: University of Chicago Press.

[56] L.J. Hubert and P. Arabie (1985) Comparing partitions, *Journal of Classification, 2*, 193-218.

[57] P. Jaccard (1908) Nouvelles recherches sur la distribution florale, *Bulletine de la Société Vaudoise de Sciences Naturelles*, 44, 223-370.

[58] A.K. Jain and R.C. Dubes (1988) *Algorithms for Clustering Data*, Englewood Cliffs, NJ: Prentice Hall.

[59] A.K. Jain, M.N. Murty, and P.J. Flynn (1999) Data clustering: A review, *ACM Computing Surveys*, 31, n. 3, 264-323.

[60] C.V. Jawahar, P.K. Biswas, and A.K. Ray (1995) Detection of clusters of distinct geometry: A step toward generalized fuzzy clustering, *Pattern Recognition Letters, 16*, 1119-1123.

[61] I.T. Jolliffe (1986) *Principal Component Analysis.* New York: Springer-Verlag.

[62] L. Kaufman and P. Rousseeuw (1990) *Finding Groups in Data: An Introduction to Cluster Analysis*, New York: J. Wiley & Son.

[63] M.G. Kendall and A. Stuart (1979) *The Advanced Theory of Statistics, 2*, 4th ed. , New York: Hafner.

[64] M.K. Kerr and G.A. Churchill (2001) Bootstrapping cluster analysis: Assessing the reliability of conclusions from microarray experiments, *Proceedings of the National Academy of Science USA*, 98, no. 16, 8961-8965.

[65] H.A.L Kiers (1997) Weighted least squares fitting using ordinary least squares algorithms, *Psychometrika, 62*, 251-266.

[66] W. Klösgen (1996) Explora – A multipattern and multistrategy discovery assistant, In [29], 249-271.

[67] T. Kohonen (1995) *Self-Organizing Maps*, Berlin: Springer-Verlag.

[68] E. Koonin and M. Galperin (2002) *Sequence–Evolution–Function: Computational Approaches in Comparative Genomics*, Dordrecht: Kluwer Academic Publishers.

[69] B. Kovalerchuk and E. Vityaev (2000) *Data Mining in Finance: Advances in Relational and Hybrid Methods*, Boston/Dordrecht/London: Kluwer Academic Publishers.

[70] D.E Krane and M.L. Raymer (2003) *Fundamental Concepts of Bioinformatics*, San Francisco, CA: Pearson Education.

[71] W. Krzanowski and Y. Lai (1985) A criterion for determining the number of groups in a dataset using sum of squares clustering, *Biometrics, 44*, 23-34.

[72] W.J. Krzanowski and F.H.C. Marriott (1994) *Multivariate Analysis*, London: Edward Arnold.

[73] S. Laaksonen (2000) Regression-based nearest neighbour hot decking, *Computational Statistics, 15*, 65-71.

[74] G. Lakoff (1990) *Women, Fire, and Dangerous Things: What Categories Reveal About the Mind*, Chicago: University of Chicago Press.

[75] G.N. Lance and W.T. Williams (1967) A general theory of classificatory sorting strategies: 1. Hierarchical Systems, *The Computer Journal, 9*, 373-380.

[76] M.H.Law, A.K.Jain, and M.A.T. Figueirido (2003) Feature selection in mixture-based clustering, *Advances in Neural Information Processing Systems, 15.*

[77] L. Lebart, A. Morineau, and M. Piron (1995) *Statistique Exploratoire Multidimensionnelle*, Paris: Dunod.

[78] E. Levine and E. Domany (2001) Resampling method for unsupervised estimation of cluster validity, *Neural Computation, 13*, 2573-2593.

[79] R.J.A. Little and D.B. Rubin (1987) *Statistical Analysis with Missing Data*, J. Wiley & Sons.

[80] T. Margush and F.R. McMorris (1981) Consensus n-trees, *Bulletin of Mathematical Biology*, 43, 239-244.

[81] R.M. McIntyre and R.K. Blashfied (1980) A nearest-centroid technique for evaluating the minimum variance clustering procedure, *Multivariate Behavioral Research, 22*, 225-238.

[82] G. McLachlan and K. Basford (1988) *Mixture Models: Inference and Applications to Clustering*, New York: Marcel Dekker.

[83] J.B. MacQueen (1967) Some methods for classification and analysis of multivariate observations, L. Lecam and J. Neymen (Eds.) *Proceedings of 5th Berkeley Symposium, 2*, 281-297, University of California Press, Berkeley.

[84] G.W. Milligan (1981) A Monte-Carlo study of thirty internal criterion measures for cluster analysis, *Psychometrika*, 46, 187-199.

[85] G.W. Milligan (1989) A validation study of a variable weighting algorithm for cluster analysis, *Journal of Classification*, 6, 53-71.

[86] G.W. Milligan and M.C. Cooper (1985) An examination of procedures for determining the number of clusters in a data set, *Psychometrika*, 50, 159-179.

[87] G.W. Milligan and M.C. Cooper (1988) A study of standardization of the variables in cluster analysis, *Journal of Classification, 5*, 181-204.

[88] B. Mirkin (1987) Additive clustering and qualitative factor analysis methods for similarity matrices, *Journal of Classification, 4*, 7-31; Erratum (1989), *6*, 271-272.

[89] B. Mirkin (1990) Sequential fitting procedures for linear data aggregation model, *Journal of Classification, 7*, 167-195.

[90] B. Mirkin (1996) *Mathematical Classification and Clustering*, Dordrecht: Kluwer Academic Press.

[91] B. Mirkin (1997) L_1 and L_2 approximation clustering for mixed data: scatter decompositions and algorithms, in Y. Dodge (Ed.) L_1*-Statistical Procedures and Related Topics*, Hayward, Ca.: Institute of Mathematical Statistics (Lecture Notes-Monograph Series), 473-486.

[92] B. Mirkin (1999) Concept learning and feature selection based on square-error clustering, *Machine Learning, 35*, 25-40.

[93] B. Mirkin (2001) Eleven ways to look at the chi-squared coefficient for contingency tables, *The American Statistician*, 55, no. 2, 111-120.

[94] B. Mirkin (2001) Reinterpreting the category utility function, *Machine Learning*, 45, 219-228.

[95] B. Mirkin and R. Brooks (2003) A tool for comprehensively describing class-based data sets, In J.M. Rossiter and T. Martin (Eds.) *Proceedings of 2003 UK Workshop on Computational Intelligence*, University of Bristol, UK, 149-156.

[96] B. Mirkin and E. Koonin (2003) A top-down method for building genome classification trees with linear binary hierarchies, In M. Janowitz, J.-F. Lapointe, F. McMorris, B. Mirkin, and F. Roberts (Eds.) *Bioconsensus*, DIMACS Series, V. 61, Providence: AMS, 97-112.

[97] B. Mirkin, M. Levin, and E. Bakaleinik (2002) Intelligent K-Means clustering in analysis of newspaper articles on bribing, *Intelligent Systems*, Donetsk, Ukraine, n. 2, 224-230.

[98] B. Mirkin and I. Muchnik (2002) Layered clusters of tightness set functions, *Applied Mathematics Letters*, 15, 147-151.

[99] M. Mitchell (1998) *An Introduction to Genetic Algorithms (Complex Adaptive Systems)*, MIT Press.

[100] D.S. Modha and W.S. Spangler (2003) Feature weighting in k-means clustering, *Machine Learning*, 52, 217-237.

[101] S. Monti, P. Tamayo, J. Mesirov, and T. Golub (2003) Consensus clustering: A resampling-based method for class discovery and visualization of gene expression microarray data, *Machine Learning*, 52, 91-118.

[102] J. Mullat, Extremal subsystems of monotone systems: I, II, *Automation and Remote Control*, 37, 758-766, 1286-1294 (1976).

[103] F. Murtagh (1985) *Multidimensional Clustering Algorithms*, Heidelberg: Physica-Verlag.

[104] S. Nascimento, B. Mirkin, and F. Moura-Pires (2003) Modeling proportional membership in fuzzy clustering, *IEEE Transactions on Fuzzy Systems*, 11, no. 2, 173-186.

[105] Nei, M. and Kumar, S. (2000) *Molecular Evolution and Phylogenetics*, Oxford University Press.

[106] K. Pearson (1901) On lines and planes of closest to systems of points in space, *The London, Edinburgh and Dublin Philosophical Magazine and Journal of Science*, Sixth Series 2, 559-572.

[107] M. Perkowitz and O. Etzioni (2000) Towards adaptive Web sites: Conceptual framework and case study, *Artificial Intelligence, 118*, 245-275.

[108] S.M. Perlmutter, P.C. Cosman, C.-W. Tseng, R.A. Olshen, R.M. Gray, K.C.P.Li, and C.J. Bergin (1998) Medical image compression and vector quantization, *Statistical Science*, 13, 30-53.

[109] K.S. Pollard and M.J. van der Laan (2002) A method to identify significant clusters in gene expression data, *U.C. Berkeley Division of Biostatistics Working Paper Series*, 107.

[110] J. Qin, D.P. Lewis, and W.S. Noble (2003) Kernel hierarchical gene clustering from microarray expression data, *Bioinformatics*, 19, 2097-2104.

[111] J.R. Quinlan (1993) *C4.5: Programs for Machine Learning*, San Mateo: Morgan Kaufmann.

[112] W.M. Rand (1971) Objective criteria for the evaluation of clustering methods, *Journal of American Statistical Association, 66*, 846-850.

[113] Repository of databases for machine learning and data mining, Urvine, UCL.

[114] R.J. Roiger and M.W. Geatz (2003) *Data Mining: A Tutorial-Based Primer*, Addison Wesley, Pearson Education, Inc.

[115] Romesburg, C.H (1984) *Cluster Analysis for Researchers*, Belmont, Ca: Lifetime Learning Applications. Reproduced by Lulu Press, North Carolina, 2004.

[116] G. Salton (1989) *Automatic Text Processing: The Transformation, Analysis, and Retrieval of Information by Computer*, Addison-Wesley.

[117] S. L. Salzberg (1998) Decision trees and Markov chains for gene finding, In S.L. Salzberg, D.B. Searls & S. Kasif (Eds.) *Computational Methods in Molecular Biology*, 187-203, Amsterdam, Elsevier Science B.V.

[118] J.L. Schafer (1997) *Analysis of Incomplete Multivariate Data*, Chapman and Hall.

[119] *SAS/ETS User's Guide*, Version 8 (2000) Volumes 1 and 2, SAS Publishing.

[120] C. Seidman (2001) *Data Mining with Microsoft SQL Server 2000 Technical Reference*, Microsoft Corporation.

[121] R.N. Shepard and P. Arabie (1979) Additive clustering: representation of similarities as combinations of overlapping properties, *Psychological Review, 86*, 87-123.

[122] P.H.A. Sneath and R.R. Sokal (1973) *Numerical Taxonomy*, San Francisco: W.H. Freeman.

[123] M. Sonka, V. Hlavac, and R. Boyle (1999) *Image Processing, Analysis and Machine Vision*, Pacific Grove, Ca: Brooks/Cole Publishing Company.

[124] J.A. Sonquist, E.L. Baker, and J.N. Morgan (1973) *Searching for Structure*, Institute for Social Research, Ann Arbor: University of Michigan.

[125] R. Spence (2001) *Information Visualization*, ACM Press/Addison-Wesley.

[126] R. Srikant, Q. Vu, and R. Agraval (1997) Mining association rules with the item constraints, in (Eds. D. Heckerman, H. Manilla, D. Pregibon & R. Uthrusamy) *Proceedings of Third International Conference on Knowledge Discovery and Data Mining*, Menlo Park, CA, AAAI Press, 67-73.

[127] M. Steinbach, G. Karypis, and V. Kumar (2000) A comparison of document clustering techniques, In *Proc. Workshop on Text Mining, 6th ACM SIGKDD Int. Conf. on Knowledge Discovery and Data Mining*, www-users.cs.umn.edu/ karypis/ publications/Papers/PDF/doccluster.pdf.

[128] C.A. Sugar and G.M. James (2003) Finding the number of clusters in a data set: An information-theoretic approach, Journal of the American Statistical Association, 98, n. 463, 750-778.

[129] R. Tibshirani, G. Walther, and T. Hastie (2001) Estimating the number of clusters in a dataset via the Gap statistics, *Journal of the Royal Statistical Society B*, 63, 411-423.

[130] O. Troyanskaya, M. Cantor, G. Sherlock, P. Brown, T. Hastie, R. Hastie, R. Tibshirani, D. Botsein, and R.B. Altman (2001) Missing value estimation methods for DNA microarrays, *Bioinformatics 17*, 520-525.

[131] R.C. Tryon (1939) *Cluster Analysis*, Ann Arbor: Edwards Bros.

[132] C. Tsallis, R.S. Mendes, and A.R. Plastino (1998) The role of constraints within generalized nonextensive statistics, *Physica A, 261*, 534-554.

[133] M. A. Turk and A. P. Pentland (1991) Eigenfaces for recognition, *Journal of Cognitive Neuroscience, 3*, 71-96.

[134] Vivisimo Document Clustering Engine (2003) http://vivisimo.com.

[135] J.H. Ward, Jr (1963) Hierarchical grouping to optimize an objective function, *Journal of the American Statistical Association, 58*, 236-244.

[136] I. Wasito and B. Mirkin (2005) Nearest neighbour approach in the least-squares data imputation algorithms, *Information Systems, 169*, 1-25.

[137] A. Webb (2002) *Statistical Pattern Recognition*, Chichester, England: J. Wiley & Sons.

[138] A. Weingessel, E. Dimitriadou, and S. Dolnicar (1999) An examination of indexes for determining the number of clusters in binary data sets, Working Paper No. 29, Vienna University of Economics, Wien, Austria.

[139] S.M. Weiss, N. Indurkhya, T. Zhang, and F.J. Damerau (2005) *Text Mining: Predictive Methods for Analyzing Unstructured Information*, Springer Science+Business Media.

[140] D. Wishart (1999) *The ClustanGraphics Primer*, Edinburgh: Clustan Limited.

[141] K.Y. Yeung, C. Fraley, A. Murua, A.E. Raftery, and W.L. Ruzzo (2001) Model-based clustering and data transformations for gene expression data, *Bioinformatics, 17*, no. 10, 977-987.

[142] S. Zhong and J. Ghosh (2003) A unified framework for model-based clustering, *Journal of Machine Learning Research*, 4, 1001-1037.

Index